Advance Praise for *Hell or High Water*

"Hell or Hig̶————————————————————————̶ ̶icanes and
folkways. But it ъ .̶ ̶, it's a mystery: Why would
people so faithfully return to this imperiled and lethal landscape?
Thibodeaux, a fine journalist and masterful raconteur, persuasively
solves that mystery."
–Jed Horne, author of *Breach of Faith: Hurricane Katrina
and the Near Death of a Great American City*

"When people or societies come under intense pressure, they either
fall apart or come together. This is the story of people coming
together, people determined to not just survive but to make things
whole."
–John M. Barry, author of *Rising Tide: The Great Mississippi Flood
of 1927 and How It Changed America* and *Roger Williams and the
Creation of the American Soul: Church, State, and the Birth of Liberty*

"Ron Thibodeaux provides us with another stirring chapter in the
long history of Cajun determination, told through the stories of
ordinary people who decided, in the words of the old Acadian folk
tale, 'j'avons pleuré assez, c'est l'heure de coupe: du bois' (we've
cried long enough, it's time to cut wood)."
–John Mack Faragher, author of *A Great and Noble Scheme:
The Tragic Story of the Expulsion of the French Acadians
from their American Homeland*

"Ron Thibodeaux trains his considerable journalistic talents and
his insider knowledge on two overlooked hurricanes that roared
into his beloved home state in the giant shadow of Katrina. . . .
Thibodeaux brings writerly flair and deft narrative to these largely
untold stories of a resilient folk whose very way of life is under
siege. As such, *Hell or High Water* is a hugely important addition to
the canon of the literature of South Louisiana's imperiled coast and
the Cajuns whose lives and culture are inseparable from it."
–Ken Wells, author of *The Good Pirates of the Forgotten Bayous*

"Ron Thibodeaux's *Hell or High Water* weaves Gulf Coast history into the true modern account of neighbor helping neighbor and parish helping parish. As told through his well-chosen words and journalistic style, the recent 'forgotten' hurricanes echo pre-24 hour news storms, when the people, without national attention or assistance, took care of themselves."

–George Rodrigue, Artist

"Journalist Ron Thibodeaux returns to his native Cajun country to tell a story of survival that is engaging, informative, and compassionate. Their losses are staggering, but the grit, humor and doggone determination of these resilient people is deeply moving— and offers lessons and inspiration for the rest of us."

–Amy Dickinson, "Ask Amy" advice columnist and NPR contributor

"Ron Thibodeaux's book is a gumbo of colorful Cajun characters, first-rate writing, and spicy cautionary tales. It's a must read for anyone who wants to understand 'hurricane culture' in this much-loved and long-suffering land."

–Mike Tidwell, author of *Bayou Farewell: The Rich Life and Tragic Death of Louisiana's Cajun Coast*

Hell or High Water

Hell or High Water

How Cajun Fortitude Withstood Hurricanes Rita and Ike

Ron Thibodeaux

Foreword by James Carville

UNIVERSITY OF LOUISIANA AT LAFAYETTE PRESS
2012

Passages from the following chapters were orginally printed in the *The Times-Picayune* and are reprinted with permission:
Chapter 2 "We Know the Drill" © 2007 by *The Times-Picayune*
Chapter 3 "The Acadian Connection" © 2004 by *The Times-Picayune*
Chapter 12 "Home Is What You Make It" © 2007 by *The Times-Picayune*

Printed on acid-free paper.

ISBN 13 (paper): 978-1-935754-11-4

Library of Congress Cataloging-in-Publication Data

Thibodeaux, Ron, 1958-
  Hell or high water : how Cajun fortitude withstood hurricanes Rita and Ike / Ron Thibodeaux ; foreword by James Carville.
     p. cm.
  Includes index.
  ISBN 978-1-935754-11-4 (pbk. : alk. paper)
 1. Emergency management--Louisiana. 2. Hurricane Rita, 2005. 3. Hurricane Ike, 2008. 4. Cajuns. I. Title.
  HV551.4.L8T55 2012
  363.34'92209763--dc23

                              2012001809

University of Louisiana at Lafayette Press
P.O. Box 40831
Lafayette, LA 70504-0831
http://ulpress.org

With love to Robyn,
who could have been a writer
but settled for marrying one instead

Table of Contents

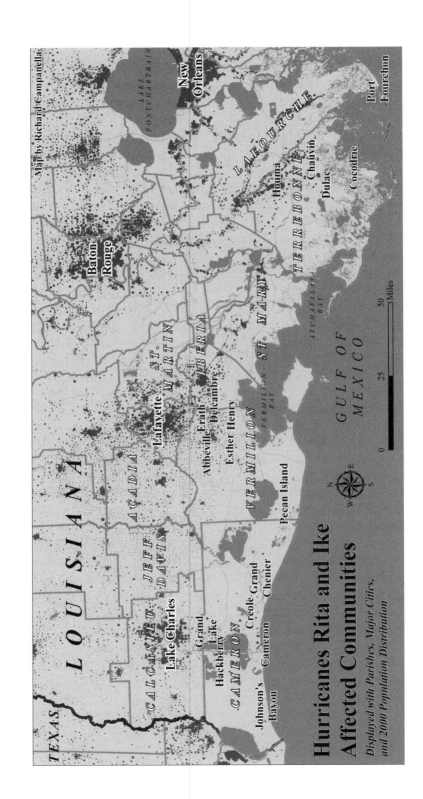

Hurricanes Rita and Ike
Affected Communities

Displayed with Parishes, Major Cities,
and 2000 Population Distribution

FOREWORD

Louisiana is not like any other place in the world. It can be said that New Orleans is the most interesting city in the United States—but it's not the only place that makes Louisiana unique. The "Cajun country" of South Louisiana is distinctive from New Orleans, the rest of the state, and the rest of the South, and it is an integral part of Louisiana's character. For the Cajuns and Creoles and their neighbors in Acadiana, life on the bayous, in the swamps and marshes, and across the rural prairie is flavored by its own food, musical styles, French accents, history, and traditions—born of a modest, long-suffering people and now recognized the world over.

When Katrina struck in 2005, the world's attention focused on New Orleans, and with good reason. From the "federal flood" to the dramatic rooftop rescues to the human misery of the Superdome and the Convention Center to the torturously slow recovery to the finger-pointing at every level of officialdom, the crisis was a major news story as the future of one of America's most influential cities hung in the balance.

Against that backdrop, it should not have been surprising that scant attention was paid to Hurricane Rita, which hammered the entire coastal region of South Louisiana just three and a half weeks after Katrina made landfall. Because Rita bypassed the New Orleans area, its impact on the rest of Louisiana was largely overlooked by the national media and, by extension, most Americans. It was a major hurricane that delivered a major blow to cities and towns all across the lower reaches of our state, though, and the people who lost their homes, churches, schools, and in some places entire communities were left to pick up the pieces themselves, out of the public eye.

Three years later, Hurricane Ike did the same thing, barreling across the Louisiana coast from end to end and flooding those Cajun communities all over again.

How the hardy people of the bayous overcame that adversity, time and again, is a story that is not well known—but it's one worth telling. The story provides only the latest chapter in the rich and unique history of Louisiana's Cajun people. The survival and endurance of those people and their culture, distinctive as it is, make for a quintessentially American story.

Ron Thibodeaux, veteran editor at *The Times-Picayune* and the newspaper's resident expert on Cajun culture, is uniquely positioned to tell this story. It is with great detail that he leads readers through the hard times of hurricane season in South Louisiana and brings forth a tale of perseverance and stubborn strength unique to our state.

On a personal note, I graduated from Ascension Catholic High School in Donaldsonville, Louisiana, and our athletic competitors were in places like Houma, Golden Meadow, and Larose—all of which have been profoundly affected by the kinds of things that Ron is writing about. This is a singular, productive, and important part of our country that has long been neglected, and this work should help to convey the real story about the people of South Louisiana and the place they call home.

– *James Carville*
New Orleans, Louisiana

PREFACE

The unique rhythms of life in Louisiana are governed by the seasons. Mardi Gras season, which lasts only a few weeks, coincides with the arrival of crawfish season, which extends for much of the year. Football season is a harbinger of duck season. Planting season, festival season, and shrimping season all have familiar homes on the cultural calendar.

When it comes to commanding respect, instilling dread, disrupting lives and livelihoods, and holding sway over entire communities, though, one time of year overshadows all others. From June to November, year in and year out, we live with one eye on the weather report, because it's hurricane season.

Tradition, family history, and personal experience teach us that major hurricanes from the Gulf of Mexico are the most destructive force known to Louisiana, furies of wind and water capable of wiping out, across a broad expanse of our state, not just homes and businesses but entire communities, while even those of lesser intensity can cause substantial damage and upheaval. Scientific advances have allowed us to better prepare for these weather phenomena, but we cannot combat them. The hurricane remains an irresistible force to be avoided—or endured—but it is not to be ignored.

Over the past six decades, the worst of them have lived on in our collective memory as entities with distinct personalities. A hurricane in 1947 killed more than fifty people in the New Orleans area, but in the litany of Louisiana natural disasters it requires a detailed description ("the storm of 1947 that flooded Slidell and Metairie . . .") just to rise above footnote status. Since the National Weather Service began assigning names to tropical storms in 1953, however, any really bad hurricane has been assured its place in history. Audrey killed hundreds of people in 1957 and, more than fifty years later, it remains a pivot point of everyday life in Cameron Parish. To this day, its very name conveys an accounting of shattered lives and battered towns, grief and recovery. You only need to say "Audrey"; no further description is necessary.

Betsy, in 1965, had the same effect in the opposite coastal corner of the state. For forty years it was the killer storm of southeastern Louisiana by which others were measured. Like Audrey, its name has evoked emotions and memories about loss and devastation that have passed on from one generation to the next.

History was rewritten, though, in 2005.

Long after Hurricane Katrina struck the New Orleans area on August 29 of that year, its impact remained a major, national story. People everywhere were spellbound by the long arc of ruin and recovery in America's most interesting city: the loss of life, the failure of the levees, the inept government response to the flooding, the police corruption, the disputes over how—and whether—to rebuild, the repair of the Superdome, and the revival of the city's restaurant and music scenes.

The crisis captured not only the public's attention, but also its imagination. Books were written about Katrina—accounts of tragedy and heroism, memoirs by survivors and scoundrels, ruminations on what was, what had become, and what could be. Spike Lee made a movie. More books were published. Politicians and pundits continued to pontificate and ridicule. More books were published. "K-Ville," a crime drama set in post-Katrina New Orleans, hit prime time. And more books were published.

As an editor at the daily newspaper in New Orleans, I had a panoramic view of the spectacle. But as the months went by and 2005 turned to 2006, and then to 2007, I kept wondering, "What about Rita?"

Just three and a half weeks after Katrina struck the New Orleans area, Hurricane Rita swept across South Louisiana, pounding the coastal communities of Cajun country before making landfall near the state's western border. Loss of life was minimal in Louisiana, but the small towns and winding bayou communities across the entire base of the state's Acadiana triangle got clobbered. Schools, churches, grocery stores, and other local institutions were destroyed; in some places, entire neighborhoods were washed away.

The damage stretched across a 250-mile swath of Louisiana's coastal parishes, but beyond the impact zone, hardly anyone knew about it. When Rita hit, it got everyone's attention . . . for about a day, it seemed. Then the national media and federal authorities and everyone else—perhaps unimpressed with the low body count, perhaps dissuaded by the relative inconvenience of uncelebrated places like Cameron, Franklin, and Abbeville as compared to New Orleans—turned their attention back to Katrina's aftermath.

For those up and down the bayous and across the marshes, the hardships inflicted by Rita were extreme. Undaunted, many residents got about the business of recovery straightaway, doing for themselves as they had always done, helping their neighbors when they could. This response reflected the best traditions of South Louisiana's rural communities, but it was happening in the blind spot of Katrina recovery.

It caught my attention, though. Born on the bayou and reared in a family where the old folks still spoke French—at least to each other, if not to the children—I have always taken an interest in my Cajun/Acadian background and the ways in which the Acadiana region distinguishes itself from the rest of our state and, indeed, the nation. As a longtime editor and staff writer at *The Times-Picayune*, I had found occasional opportunities over the years to write meaningful stories about Louisiana's Cajun culture for the readers of our newspaper in the not-very-Cajun New Orleans area. Now, from my vantage point in the suburbs of the big city, I sensed that this was a story deserving of a wider audience. Convinced by the spring of 2007 that it wasn't going to get done if I didn't do it myself, I began setting off in my spare time for places like Grand Chenier, Erath, and Dulac, in search of the people whose hurricane experiences would provide the foundation for this book.

As I chipped away at research and interviews for the project, Hurricane Ike came along in 2008 and, incredibly, followed Rita's script. It was a huge, powerful hurricane. It swept from east to west across the upper Gulf of Mexico. It swamped most of the same communities from one side of coastal Louisiana to the other. And, true to form, it got scant attention from the outside world, as the national media focused on New Orleans, which avoided significant trouble, and Galveston, Texas, where the hurricane made landfall.

The flooding actually was worse for Ike than for Rita in some parts of South Louisiana, including the bayou communities south of Houma. From Terrebonne in the east to Cameron in the west, newly renovated homes, churches, and businesses were once again ruined, leaving weary residents to pick up the pieces once more. That's exactly what they did—and that became part of the story.

Thus, my Hurricane Rita book became a Rita-and-Ike book.

Inanimate though they were, Rita and Ike—with their human names and menacing personalities—are important characters of this narrative. Mostly, though, this is a story about the people of South Louisiana, their devotion to the place they call home, and their fortitude in the face of substantial adversity. For those who don't know, or don't know enough, about what Hurricane Rita and Hurricane Ike did to Louisiana, I hope the story is enlightening. And for those determined individuals who withstood the storms, held their communities together and, in doing so, sustained their way of life for Cajun country's future generations, I hope my telling of their experiences does them justice.

– Ron Thibodeaux

STORM TIDE ON THE BAYOU

When the refrigerator started floating across the kitchen, Adam Suire knew it was time to get out of the house.

A tour of duty on the open seas as a merchant sailor and thirty years as an oil field roughneck had taught the Louisiana native to respect the water, but never had he feared it. At seventy-seven, he was still as tough—and stubborn—as he had ever been, so when he failed to heed multiple warnings from local officials to evacuate ahead of the approaching hurricane, no one was particularly surprised. After all, he had stayed put for all the hurricanes that had blown through in the previous forty years, and none of those had ever flooded his house. Why should this one be any different?

The eye of the storm had already churned past, and residents like Suire, his wife, and their daughter and son-in-law who had chosen to ride it out thought the worst was over. Hours later, though, the water started rising. It overtook yards up and down the street, then began to engulf the houses. There was no keeping it out, and all of a sudden it was too late to leave. As furniture began to slosh around, the family climbed into the attic. Before long, the floodwaters found them there and chased them out to the roof.

With the onslaught threatening to either break the house apart or swallow it whole, they retreated from the rooftop and scrambled up what they hoped would be a more secure perch: an oak tree. There they held on for hours, soaked by the rushing tide, battered by debris that the current and the winds aimed at them, threatened by water moccasins. The flooding continued to spread, and local authorities and volunteers set out in a flotilla of airboats, skiffs, and other shallow-draft vessels to rescue trapped victims. The weary members of the Suire family were among the lucky ones, plucked from the tree that afternoon by a Coast Guard helicopter and brought to safety at a shelter where they got medical attention and a hot meal. Before long, conditions worsened and authorities had to suspend the rescue operations, leaving other survivors stranded until the next morning.

As Coast Guard rescuers hauled him into the HH-65 Dolphin chopper and away from the treacherous open water enveloping his neighborhood on that fateful day in 2005, Adam Suire had no way of knowing it, but he had just stared down one of the worst natural disasters ever to strike the United States.

Of course, 2005 will long be remembered as the year of the hurricane, when so many storms formed in the north Atlantic that the National Oceanic and Atmospheric Administration ran out of names and had to use letters of the Greek alphabet to identify the last six storms of the year. By the time Hurricane Zeta played out, that ruinous season had spawned seven major storms, inflicting death and destruction all across the Gulf Coast of the United States.

In the midst of all that turmoil, though, Adam Suire's storm was one of singular ferocity. It was, for a time, the largest hurricane ever measured within the Gulf of Mexico. It was one of the strongest Category 5 hurricanes ever recorded across the entire Atlantic basin. A crew of "hurricane hunters" from the Air Force Reserve clocked one of its wind gusts at 235 mph. Its approach precipitated the largest evacuation in American history. It was a killer storm, responsible for deaths across four states. When it struck South Louisiana, with an enormous tidal surge and intense winds, it threatened more than just tens of thousands of local residents. It also took dead aim at a culture and a way of life unique to all of the United States as well.

But this is not the story that most Americans think they know. This was not Hurricane Katrina, which held the country in rapt attention as the human misery throughout flooded New Orleans, compounded by inept government response at every level, played out live, around the clock, on network news telecasts.

No, this was Hurricane Rita, the *other* Louisiana disaster of 2005.

Sweeping across the upper Gulf of Mexico en route to a September 24 landfall near the state's western border, less than four weeks after Katrina hit the New Orleans area, Rita clobbered com-

Tracy Johnson of New Orleans evacuated to Hackberry for Hurricane Katrina. Two days before Rita struck, he was on the move again, crossing the Calcasieu Ship Channel from Holly Beach to Cameron aboard the ferry *Cameron II*.

munities across the entire 250-mile coastal foundation of Acadiana, America's one-of-a-kind Cajun country. From one end of the Louisiana coast to the other, towns were flooded, populations were left homeless and without public services, and communities were all but wiped off the map.

As soon as Rita trailed off the National Hurricane Center's radar, it also faded from the American consciousness. The destruction and rebuilding of New Orleans remained a major national story, spawning books and movies, high-profile benefit events and Congressional inquiries, even a television series or two, but the bayou country communities hit so hard by Rita were all but forgotten and largely left to fend for themselves. From Contraband Bayou to Bayou Carlin to Bayou Terrebonne, members of this diverse population did what their forebears had done for centuries before them: survive, adapt, and thrive in hostile environments. Their self-sufficient response to Hurricane Rita contrasted dramatically with the apparent paralysis that simultaneously afflicted much of post-Katrina New Orleans.

Incredibly, three years later, powerful Hurricane Ike took an eerily similar track across the northern Gulf of Mexico and slammed the same communities across the entire Louisiana coast with high winds and tidal flooding as it made for Galveston, Texas. Some fared better in the 2008 storm, others worse. Once again, most of them set about rebuilding on the land their families had called home for many generations.

Major hurricanes have assaulted South Louisiana before, but Rita and Ike did what others before them had not. From far out in the Gulf of Mexico, these storms took massive swipes at the entire Louisiana shore, from east to west, rather than coming "up" from the Gulf to impact one specific location. While the rest of the country viewed Rita and Ike through the skewed perspectives of how they affected other places—New Orleans, Houston, and Galveston—the fury these storms unleashed on Louisiana's smaller coastal communities, out of America's line of vision, dropped them to their knees and left them reeling.

It all began on September 7, 2005, nine days after Katrina's landfall. As law enforcement officers and military personnel were going house-to-house in New Orleans to flush out the final ten thousand residents who had refused to leave that still-flooded, broken city, a tropical wave formed off the western coast of Africa. Ten days later, after crossing the Atlantic Ocean and converging with the remnants of a cool front north of Puerto Rico, it became a tropical depression, and the next day it was christened Tropical Storm Rita.

Rita intensified from a tropical storm to a Category 5 hurricane within thirty-six hours as it entered the Gulf of Mexico and flourished on the warm waters of the Loop Current. When it passed the Florida Keys, it was pointed in a westerly direction, toward the Texas/Mexico border. The first hurricane watch issued by the National Hurricane Center covered most of the Texas coast and extended eastward as far as Cameron, some twenty-five miles past the state line into Louisiana.

About four hundred miles south of Fort Walton Beach, Florida, the hurricane began a long, gradual shift toward the north, and the weather experts started nudging their dreaded "cone of uncertainty" further and further eastward into Louisiana. Fifty-eight hours after the initial watch advisory, Rita arrived at the far western edge of Cameron Parish as a large Category 3 hurricane.

Rita was so strong that it knocked out most of the weather buoys in its path hours before it made landfall, and almost all the homes and buildings in the communities stretched out along the coast road, Louisiana 82, succumbed to the ravages of the hurricane's most intense section, the northeast quadrant. Consequently, with few high-water marks to study, no eyewitness accounts due to an effective evacuation, and the lack of data recorded by weather gauges as the storm approached, the National Hurricane Center struggled to calculate the storm's precise characteristics at landfall. Its best guess gave Rita sustained winds of at least 120 mph as it came ashore, whipping up massive waves atop a storm surge of fifteen feet or more.

As Rita made its way through the Gulf, most of the nation's attention was fixated on the prospects that it could do further harm to New Orleans, just three and a half weeks after Katrina. From hundreds of miles away, Rita's wave action did push into the city's Lower Ninth Ward, but the practical impact proved to be minimal since no meaningful repairs from Katrina's widespread flooding there had yet begun. Meanwhile, in Texas, a mandatory evacuation was called for the metropolitan Houston area, well ahead of Rita's landfall. As many as two million people fled, jamming the region's freeways for forty-eight hours. The action was hailed as the largest evacuation in U.S. history. The exercise contributed to the deaths of many elderly and otherwise sickly people who suffered debilitating effects of the massive traffic jams in southeast Texas' ninety-eight degree heat, however, as thousands of frustrated, gridlocked motorists were forced to turn off their air conditioners to conserve gasoline.

There was just one Rita-related death reported in Louisiana, a drowning in Calcasieu Parish, but the storm experience proved to be a monumental collective challenge for the people of South Louisiana.

Many who evacuated Cameron Parish could not return to their homes in a few days, weeks, or even months, for they had no more homes, no more hometowns. Holly Beach was washed away in its entirety; Johnson's Bayou, nearly so. The fortress-like courthouse remained intact in nearby Cameron, but virtually every home and business was destroyed. The parish's vast marshes were littered with furniture, clothing, appliances, and bits and pieces of the houses that used to contain them, along with plenty of boats and dead cows and a few hundred coffins that the floodwaters had extracted from burial sites and mausoleum crypts.

From Cameron, the surge pushed more than thirty miles up the Calcasieu River, flooding downtown Lake Charles, destroying Harrah's dockside casino, and badly damaging the lakefront civic center. The powerful hurricane-force winds ripped roofs off some houses and sent trees crashing onto many others. South of the city, Lake Charles' airport terminal was wrecked by a tornado.

In lower Vermilion Parish, 125 miles east of Rita's landfall, the force of the storm surge smashed some homes apart on impact, washed some houses miles away, and flooded others up to the roof line. Unlike in neighboring Cameron Parish, where the locals remembered and respected the lessons of a benchmark hurricane from almost a half-century past, hundreds of people in vulnerable Vermilion Parish communities such as Forked Island, Henry, and Delcambre ignored orders to evacuate. Many of them, like the Suires on Jean Marie Road south of Erath, were forced out of their homes by floodwaters and had to be rescued by sheriff's deputies, state wildlife agents, and local volunteers in boats or Coast Guard personnel in helicopters.

The storm surge that wiped out lower Vermilion Parish not only inundated the low-lying region abutting the Gulf and Vermilion Bay, but it also flooded cow pastures, rice farms, and other open land extending north toward Abbeville. One stranded community was Pecan Island, a settlement on one of the narrow oak-lined ridges known as cheniers, rising slightly above the southwestern Louisiana coastal marshes. Until help arrived, residents who had stayed behind there were marooned on an actual island, cut off from the rest of the parish by thirty miles of open water.

One hundred miles east, the torrent cut through everything in the isolated bayou communities of lower Terrebonne Parish. Homes, stores, and churches that weren't lost to the flood were left with a deep, smelly coating of marsh mud once the water receded. Nasty, toxic mold soon would follow.

Even after the flooding dissipated, the authorities locked down many areas wracked by Rita. With no electricity, no water, no other utilities available at first, and in many locations no businesses open, officials saw little point in letting people return to their homes, or what was left of them. But residents were eager to get in, assess their damage, and start making repairs if repairs were possible. Some didn't even wait for the water to go down, heading out by boat and skirting official checkpoints.

In the case of lower Cameron Parish, repairs were seldom possible because most homes had been washed away, so the options ran more toward setting up a mobile home or camper and deciding whether and how to build another home on-site. "I would use the word destroyed," Army Lt. Gen. Russel Honoré, commander of U.S. military relief operations in Louisiana, said of Cameron and nearby Creole at a press briefing after his initial visit to southwestern Louisiana. "Most of the houses and public buildings no longer exist or are even in the same location that they were."

Since New Orleans to one side and Houston and Galveston to the other managed to avoid major hits, the national media, and thus the nation at large, quickly lost interest in the Hurricane Rita story. In the heart of the impact zone, it didn't take long for tens of thousands of beleaguered storm victims across coastal Louisiana to recognize that for all the destruction it caused, Hurricane Rita was quickly becoming "the forgotten storm" as the country's attention turned back to New Orleans.

Several circumstances contributed to this Rita amnesia.

The devastation of New Orleans by Hurricane Katrina was not only a major, international story—it also was a fresh one. Barely a month after Katrina made landfall, significant developments in New Orleans continued to unfold day by day. With national media already in place, the flooding, death, chaos, dramatic rescues, human suffering, political intrigue, and slow return to order in a major American city, especially one with as much character and historical significance as New Orleans, made for gripping television.

On the other hand, Hurricane Rita, for all its destructive force, hit a less populated area, one that had largely evacuated ahead of the storm. Its impact was easily overlooked by network television news teams who had positioned themselves at the major cities—New Orleans, Galveston, and Houston. Katrina killed 1,577 people from Louisiana; Rita killed one. Afterward, resourceful storm victims in one coastal community after another made their way back home and, without fanfare, began putting their lives back together.

"Knowing these people, most of them are hunters, trappers, farmers," Robert LeBlanc, who was Vermilion Parish's director of emergency preparedness at the time, told reporters the day after Rita struck, when some evacuees already were getting antsy about wanting to check out their homes. "They're not going to wait on FEMA or anyone else. They're going to do what they need to do. They're used to primitive conditions."

Without being asked, men and boys fanned out on horseback to help neighbors round up stray cattle, in some cases before the floodwaters had even gone down. In neighborhoods throughout the industrial, college town of Lake Charles, folks who rode out the storm at home were joined by others who had evacuated but quickly got back by avoiding roadblocks that authorities set up at major entry points; together, with their own chain saws and winch-equipped pickup trucks and sense of community, they began clearing fallen trees from the streets of their neighborhoods and the houses of their neighbors. In Abbeville, people like Liz Touchet and Diane Meaux Broussard, the parish clerk of court, jumped right in to organize relief efforts to assist local residents with food, supplies, and other needs.

Such actions were ringing endorsements of can-do American spirit, but they weren't going to drive ratings for network news and talk shows like the multi-faceted catastrophe taking place right down the road in New Orleans.

Three years later, Hurricane Ike did the same thing as Rita—building into a huge, strong storm, then raging east-to-west across the entire Louisiana coast. Once again, the parishes along the base of Louisiana's Acadiana region were beaten down by a fierce and powerful hurricane as it swung past. Some towns were hit almost as badly as they had been for Rita; for others, the flooding was even worse.

In Lake Charles, the water reached eleven feet in some parts of downtown, exceeding the flooding that Rita caused there. The surge pushing up from the Gulf, through Cameron Parish, swelled the Calcasieu River to its highest recorded level in ninety-five years and flooded more than 1,500 Lake Charles-area homes, some of them thirty miles inland.

Two hundred miles away in lower Terrebonne Parish, the flooding brought by Ike was the worst anyone could remember.

All across coastal Louisiana, repairs to churches, schools, businesses, and homes that had been three years in the making were wiped out in an instant, leaving residents to start over yet again.

In some other place, observers might have been moved to ask: How would those storm victims respond? Are they even coming back? How could they?

But this was not some other place. This was South Louisiana, home to a distinctive cultural tapestry wherein abundant threads of far-flung ethnic origins had been woven together across more than three centuries to create a stalwart way of life amid the marshes, swamps, and prairie.

Its myriad influences could be traced to European immigrants—Spanish, German, Scots-Irish, Italian, and French—as well as African slaves, refugees from Santo Domingo, free people of color, Native Americans, and twentieth century arrivals from as far away as the Middle East and as close as Texas and Oklahoma. Among the smallest immigrant groups to settle in South Louisiana were Acadian exiles—the families of French-speaking pioneers who had left their mother country in the 1600s for a new life in the New World, in the region now known as the Canadian Maritimes, only to be stripped of their possessions and expelled by the British beginning in 1755. Those that eventually made their way to Louisiana perse-

Satellite photo of Hurricane Rita sweeping across the Louisiana coast, September 24, 2005.

vered in the unfamiliar territory; through intermarriage and socialization, later generations of Acadians would expand their cultural base up and down the bayous and across the rural landscape.

Over the years, there was plenty of give-and-take in the region's folkways. Southwestern Louisiana Creoles were instrumental in introducing a horse culture and are thought by some historians to have been the first American cowboys. The "gombo" cooked by African descendants was adopted and adapted by their white counterparts. The fiddle-based folk music brought to Louisiana by the Irish and the Acadians was radically impacted by a German import, the accordion.

Meanwhile, the name "Acadian" was informally Louisianaized into "Cajun." The term would come to apply not only to Acadian descendants exclusively but also their in-laws, friends, neighbors, and kindred spirits throughout South Louisiana who embraced their French accents, their cooking styles, and their proclivity for passing a good time. Thus, Schexnayders and Nunezes and Greelys ultimately were rendered as bona fide as LeBlancs and Broussards in the modern-day Cajun culture.

That assimilation also triggered the evolution of a Cajun ethos, deeding to later generations of South Louisianians a cultural inheritance devoid of ethnic restrictions. Louisiana tourism marketers like to tout the *joie de vivre* of the Cajun people—their joy of living, a let-the-good-times-roll attitude that distinguishes them from button-down types elsewhere. Underlying that superficial characterization, though, is a strength of character forged by centuries of shared hardships, from the struggle to subsist in the harsh physical and economic conditions of Louisiana's earlier years to the forced suppression of their culture by a governing elite through much of the 1900s.

These are a people of fortitude.

And as such, they manifest an inherited, emotional response to tragedy. Their antecedents, after all, include survivors of *le grand derangement*—the "great upheaval" of the Acadian exile that saw families split apart and sent away, thousands die from disease or deprivation, and survivors imprisoned or shunned in unwelcoming locales. Down through the years, they have faced natural disasters, political and societal oppression, and man-made calamities, and they have closed ranks, endured, and sustained. The gargantuan BP oil spill of 2010, for example, posed not only a physical threat to the Gulf of Mexico and coastal wetlands but a more personal threat as well: by fouling the Gulf waters, the spill also imperiled for many Cajun families a vital means of cultural transfer, for it is on the decks of shrimp boats and, afterward, in the kitchen with the day's catch where many children still learn the language and traditions of their people from their fathers and mothers and aunts and uncles.

In cities and small towns, along bayou roads and into the sparsely populated swamps and marshes, hardy local residents—who give Louisiana much of its distinctive flavor by the very lives they lead, just as generations before them have done—time and again have responded to adversity by hauling themselves and their communities back up.

For anyone familiar with the people of South Louisiana, then, the questions that followed Hurricane Rita and Hurricane Ike could only be: How soon are they coming back? How could they not?

James Lee Witt, who had been a director of the Federal Emergency Management Agency during the 1990s, was brought in as an adviser to Louisiana's state government after Katrina and Rita. Witt urged Gov. Kathleen Babineaux Blanco and the Louisiana Recovery Authority, which she created, to think creatively regarding ways to minimize future risks from storm surge and wind damage in most exposed areas. One option he and his team floated was for the federal government to buy out homes and businesses along Louisiana 82, the main road along the Gulf Coast through much of Cameron and Vermilion parishes. The consultants envisioned keeping entire communities substantially intact, and out of harm's way, by relocating them far from the coastal flood zone. New schools would be built. New churches. New neighborhoods.

The proposal looked good on paper. But missing from the consultants' actuarial calculations of loss prevention was the projected impact on the way of life for those affected by such a radical move. The balance sheets failed to consider the intrinsic value of *la terre*—the French concept, conveyed to the New World by Acadian settlers in the 1600s and sustained through four centuries of descendants, that not only does land belong to the people, but the people belong to the land as well.

As envisioned by the government's recovery consultants, tracts where the same families had lived for as many as six or seven generations would be bought out and abandoned. For people who live their lives off the marshes, the bays, and the Gulf of Mexico—some of whom count among their ancestors those who had been forced from their Acadian homeland and sent into exile by a hostile government—the prospect of another forced removal by government decree portended grave consequences. These are people who trawl the Gulf waters for shrimp. They trap in season. The great outdoors of South Louisiana are their backyards, and they fish and crab and hunt as casually as city-dwellers and suburbanites go to Walmart or hit the fast-food drive-thru. Those endeavors, both as occupations and as pastimes, go a long way toward defining the way of life bayou folk have enjoyed for many years and have passed on within each family from generation to generation. And what's done with those bounties—the communal peeling of the shrimp, the boiling of the crabs, the making of the gumbo—isn't a chore. It's a social event, a means of bringing families together, sharing traditions, and nourishing a distinctive culture.

Relocating an entire community from a floodplain to higher ground several miles away might make sense for a river basin somewhere in the Midwest, but the same approach would be ill-suited for coastal Louisiana. The feds meant well when they proposed leaving the bought-out properties available for hunting and fishing. Once the former residents were far removed, though, the damage would be done if they had to haul their boats an hour or more just to get to the water, or if they couldn't step just outside their back door and spend an hour or two catching enough crabs to boil up for dinner but instead had to decide between Hamburger Helper and a trip to Taco Bell. To be sure, some coastal residents, scared or disheartened by the damage wrought by Rita and Ike, voluntarily opted for that trade-off and moved away from coastal areas. But a plan to unilaterally uproot entire communities from lower Cameron Parish or Vermilion Parish and relocate the people thirty or more miles inland was a vastly different matter. The move might keep more homes from flooding in the next storm surge, but at what cost if it effectively killed off a way of life that characterized the residents' very culture?

As recovery efforts proceeded, local residents and officials struggled to convey to disconnected federal authorities and outside consultants just what was at stake. Tim Creswell, an emergency preparedness staffer for Vermilion Parish government, understood the dilemma as well as anyone, and he had little patience for cookie-cutter plans that ignored the real-life situation in South Louisiana.

The state legislature created the Louisiana Coastal Protection and Restoration Authority after the 2005 hurricane season, bringing together federal, state, and local agencies and interested individuals to devise a workable plan for conserving and restoring the state's coastal wetlands and barrier shorelines while also addressing hurricane protection. At the first meeting that Creswell attended, his attention was drawn to a map of Vermilion Parish on which a large section of the southern part of the parish was shaded pink. He asked an Army Corps of Engineers official what the pink area on the map signified.

"That's buy-out and relocate," the corps staffer explained.

Creswell's reaction was knee-jerk, blunt, and spot-on: "You've lost your damn mind."

WE KNOW
THE DRILL

Clifton Hebert's first job was waiting tables at the Cameron Café.

It didn't matter that he wasn't as tall standing up as his customers were sitting down. It didn't even matter that he hadn't yet learned how to write.

When folks entered the restaurant, little Clifton—barely kindergarten age—would bound over to their table and present them with a menu, a note pad, and a pencil.

"Write down what you want," he'd tell them.

Once the order was inscribed, he would dutifully deliver the pad to the kitchen, where Lena Authement was waiting. She would read the order, thank her young grandson for a job well done, and start cooking.

Working in his grandparents' café was a rite of passage not only for Hebert but also for his father and more than a dozen aunts and uncles and cousins who preceded him. Memories of that simpler time in his life might make him wistful today, but they also prod Hebert to appreciate the influence his grandparents had on him.

Cameron has never been anything but a small town, but Hebert remembers it as a thriving place during his youth in the 1960s. There were lots of oil field-related businesses. Oyster boats jammed the docks. Three pogy plants operated back then, processing steady hauls of the small, oily, quick-to-spoil fish harvested by the millions from the Gulf of Mexico and used throughout the region for bait and feed stock. People were always coming and going, and the Cameron Café was strategically located in the midst of all that activity, right across Marshall Street from the Cameron Parish Courthouse.

Lena Richard Hebert, a native of Erath, had moved to Cameron in 1946 after a divorce. She was thirty-two years old, and she had seven sons. In time, she married Jackson Authement, an oil company supply boat operator with nine daughters. Eventually, Authement left his job with Superior Oil and the couple dove headlong into the restaurant business.

The work was 5:00 a.m. to 10:00 p.m., seven days a week. The local plant and oil field trade provided a steady stream of patrons, and Lena also was enlisted to cook for the prisoners in the parish jail, earning twenty-five cents per meal from the Cameron Parish sheriff. She saw to it, too, that all of her boys worked in the restaurant. So did their girlfriends. And when they married, found places to live in Cameron, and started families of their own, the practice continued with the next generation.

"This is what we knew as being life," Hebert said.

As a teenager, Hebert drove a truck, picking up nutria meat from trappers in the marsh and delivering it to alligator farms, where the carcasses were fed to the gators. He was struck by how many of the older trappers had their grandsons with them, learning the centuries-old skills of trapping, skinning, and drying pelts—and, in a broader sense, learning that hard work is its own reward.

"Young kids to grandfathers—that's who did this, and that was their life," Hebert recalled. "You'd ask them, 'What else do you want to do?' and they'd say, 'I don't want to do nothing else. I'm going to go to work. I'm going to work hard.'"

"It comes from the grandkids working with the grandparents."

Hebert understood, even back then. He had the same affinity for his own grandparents, and they taught him the same kinds of lessons.

"When they sold the restaurant, they bought a little shrimp boat. It was just something for them to retire to and kick back. All through my high school years, I would go out with them at night and on weekends, during butterfly season, and we'd butterfly," Hebert said, referring to the practice of shrimping with large square-rigged nets, looking something like butterfly wings, draped on either side of the boat.

Jackson and Lena would pay him for his efforts, but that wasn't the main thing.

"It was more to help them," he said. "It really wasn't about the money. It was more about being out there with your grandparents and growing up and listening to their stories and their friends' stories. It's something you can't go get on the street somewhere. That mindset that they had just drove us into the same mindset. It's what you pass on."

There's something else that the people of Cameron Parish have passed on. Clifton Hebert knows it particularly well, as a native son who grew up to become the parish's first full-time director of emergency preparedness. It's a story of fear and deliverance; of loss and survival; of pain and endurance. It's the story upon which life in Cameron Parish has pivoted since June 27, 1957. It's a story named Audrey.

On a fateful day just five months after Pres. Dwight D. Eisenhower began his second term of office, and four months before the Soviets inaugurated the space age by launching the first Sputnik satellite, time stood still in Cameron Parish. With twenty-foot waves riding a twelve-foot storm surge and winds topping out at 150 mph, Hurricane Audrey swept hundreds of unsuspecting and unprepared men, women, and children to their deaths and altered the fabric of everyday life for generations of the Cameron Parish residents left behind.

Audrey became the United States' benchmark killer storm of modern times, claiming more lives than any other named hurricane until it was eclipsed by Hurricane Katrina in August 2005. But in Cameron Parish, its impact went beyond mere statistics. For decades, parish residents have found themselves in one of two groups: those who rode out Audrey and their descendants. And everyone knew, or knew of, relatives or neighbors who perished.

Subsequent to Audrey, approaching hurricanes are always taken seriously there. Forty-eight years later, when Hurricane Rita entered the Gulf of Mexico three weeks after Hurricane Katrina struck New Orleans, the experts projected a likely landfall far away at Corpus Christi, Texas, but Cameron Parish residents knew the drill: prepare for the worst, batten down the hatches, pack a few days' supplies, get out while the getting's good, then return home when the emergency is over.

This time, the worst-case scenario played out—not only for Cameron Parish, but for the other parishes along the base of the state's Acadiana triangle as well. Rita kept drifting northward as it traversed the Gulf, and after sweeping across the Louisiana coastline, it slammed ashore in Cameron Parish with a fury not seen thereabouts since 1957.

In its wake: utter destruction.

In the days, weeks, and months that followed, amid the ruin and the wreckage and the isolation, the people of Cameron Parish faced a challenge as daunting as was encountered by anyone anywhere in the Year of the Hurricane. At stake was the future of their community—and, with it, the way of life that had defined them as a people for generations.

Tucked into Louisiana's extreme southwestern corner, Cameron Parish is a curious place. Covering 1,313 square miles, it is the largest of Louisiana's sixty-four parishes, yet it is more densely populated by egrets and alligators than people, with spits of inhabitable land surrounded by broad expanses of marsh, open water, and four national wildlife refuges. Less than ten thousand people live there—not enough to fill a respectable college basketball arena—but those residents know it as heaven on earth, where nature's bounty awaits just outside the back door, and where the steady breeze coming up from the Gulf of Mexico accentuates the leisurely pace of country life.

Coastal Cameron Parish is isolated even by modern standards. Its lone east-west highway affords a tranquil two-hour drive along the "front ridge" from Vermilion Parish to the Texas line, punctuated by a ferry crossing at the Calcasieu Ship Channel. Only two roads head north through a wide swath of scenic lakes and marshes to connect it to the mainland; it takes an hour for most residents to get to downtown Lake Charles, the nearest city. Communities such as Grand Chenier and Creole are mostly quiet, rural areas, where homes share the narrow belt of available land with cattle pastures and the occasional church or store. The only semblance of hustle and bustle turns up in Cameron, seat of parish government, home to a well-established commercial seafood port and jumping-off point for businesses servicing the offshore oil industry. The ferry crossing at Cameron leads to the settlements of Holly Beach—whose array of modest beachfront homes and camps spawned its tongue-in-cheek nickname as the "Cajun Riviera"—and Johnson's Bayou, closest to the state line at Sabine Pass.

Fifty years ago, remote Cameron Parish might as well have been half a world away from anywhere. And in an era predating Doppler radar, weather satellites, and Jim Cantore, approaching hurricanes did not attract the kind of attention that modern-day residents of the Gulf South take for granted.

Memorial for unknown victims of Hurricane Audrey, behind Sacred Heart Catholic Church in Creole.

The first tropical storm of the 1957 hurricane season sprung up in the Bay of Campeche in the southwestern Gulf of Mexico on June 24, a Monday. It took a northerly course toward the Louisiana/Texas border and in two days' time became Hurricane Audrey. Some residents along the coastal reaches of Cameron Parish evacuated inland on that day, well ahead of a landfall that forecasters were predicting for the next afternoon. Some sought shelter by Wednesday night at the bunker-like parish courthouse. Others stayed put, intending to leave Thursday morning or opting to ride out the storm at home.

It's hard to know the extent to which Cameron Parish residents were informed of what was in store for them. Only a few radio stations throughout the region could reach Cameron listeners back then. Lake Charles' lone television station had begun broadcasting just three years earlier, but reception was spotty; besides, in this isolated rural area—in 1957—many folks still considered television sets a luxury they couldn't afford or didn't need.

Whatever warnings were communicated, many people chose not to evacuate because they did not expect a hurricane that powerful, that soon. After all, Audrey was a rare June storm, appearing just four weeks into the six-month hurricane season. Furthermore, the area had not experienced a serious hurricane for many years. Following a long-established custom, many in Cameron Parish gathered in extended family groups to ride out the storm together in houses scattered throughout the coastal communities, situated at or near sea level, not far from the Gulf shoreline. And even if they heard the hurricane advisory issued by the weather bureau in New Orleans at 10:00 p.m. on that Wednesday night, all it told them was that Audrey was still about two hundred miles south of Cameron, moving northward at about 10 mph, with highest sustained winds of about 100 mph. Thus, they went to bed secure in the knowledge that if conditions got really bad Thursday morning, they would still have time to evacuate.

It would prove to be a deadly mistake, for the hurricane strengthened and sped up overnight as it approached the Louisiana coast. Residents awoke to the sound of water sloshing against their houses, or inside them. By then, the storm surge already had cut off their escape routes, and they were stranded in the path of what turned out to be a monster hurricane, with winds approaching 150 mph and a ferocious tidal surge that would reach inland for twenty-five miles.

Most dwellings, from shacks to sturdy farmhouses, proved to be no match for Audrey's fury. The storm brought torrential sheets of rain that pounded away at Cameron for hours. The winds were stronger than anyone there had ever lived through or even heard of. And whatever wasn't blown to bits by Audrey's winds was inundated, broken up, or washed away by its relentless floodwaters. As houses flooded, harried residents sought refuge in their attics. That's where some of them drowned. Others were cast out into the surging tide as homes disintegrated. Lucky ones managed to cling to debris or swim to nearby trees, but even then many were lost to the pounding of the waves—and, as the ordeal went on for hours, snake bites. Many were the survivors who would be haunted by visions of their spouses, children, or parents being swept away in the flooding.

Among the lucky ones were Twila Savoie and her husband, Willard, known to everyone as "Yank." Twila worked at Dr. George Dix's office, a wood-frame building next door to the doctor's home in Creole, about four miles from the coast. It seemed to be a sturdy place, so Yank met her there and they settled in to ride out the storm with the doctor and several other people. At one point Thursday morning, as the water was rising, the doctor left to retrieve his medical bag from his house. In the few moments it took for him to enter his house, fetch the kit, and step outside again, the surge washed away his office.

The building remained intact, but water started pouring into it as waves swept it across the marsh toward the Intracoastal Canal, with the Savoies and the others still inside. They used a coat tree from the corner of the waiting room to bust a hole in the ceiling, and everyone was able to crawl through the hole into the attic, except for one elderly woman who did not survive the ordeal. The structure came to rest in the marsh and never did break apart. Once the storm subsided, Dr. Dix and Yank's father ferried the survivors to safety by pirogue.

Audrey turned out to be the only Category 4 hurricane ever to make landfall in the United States in the month of June. While it also flooded lower Vermilion Parish to the east and took sustained winds above 100 mph into Calcasieu Parish to the north, Audrey

Shadd Savoie at the materials yard of Roy Bailey Construction.

did its worst in Cameron. The official death toll was placed at 390, but that's widely acknowledged as a low-ball figure; there were individuals or entire families whose bodies were never recovered from the area's wetlands. A variety of state, federal, and local sources have estimated the fatality total between four and six hundred.

All in all, it was, for Cameron Parish, the wrong hurricane, in the wrong place, at the wrong time.

Residents of lower Cameron Parish who survived returned to bury their dead and rebuild their communities, and Hurricane Audrey's legacy filtered down within those families, from one generation to the next.

"To this day, my father will tell you that coat tree was what saved their life," said Shadd Savoie, who was born to Yank and Twila five years after Audrey struck.

Clifton Hebert could relate to that. "We grew up hearing these stories," Hebert said. "Our grandfathers went through it, or our fathers went through it, and it's been bred in us."

As the recovery from Audrey got under way, an enterprising young man came down from Ottumwa, Iowa, to get in on the action.

His name was Roy Bailey, and he arrived with a single-axle dump truck and an unflappable work ethic. He started out helping with the repair of roads that had been washed out or otherwise damaged by the flooding. An asphalt plant was set up in Oak Grove, and Bailey spent his days hauling the asphalt to the road construction sites. Every night, he would drive to Holmwood, twenty-five miles away in Calcasieu Parish, for raw materials to restock the plant.

Once the roads were completed, people eager to build up their home or business sites started asking the dump-truck driver for loads of dirt. Bailey made some inquiries and purchased a tract of land, high and dry and away from the coastal marshes, that could provide the soil that was suddenly in demand. To excavate the dirt, he bought an old, used dragline and had it delivered to his property. He had never operated a dragline before, but he figured it out in short order and by day's end he was loading his truck and hauling dirt. Roy Bailey Construction was on its way.

A friendship with one of Cameron Parish's major local contractors provided entrée for Bailey to get work in the oil fields, doing site preparation, building roads, and providing other services for inland drilling operations, and the company flourished. By the time he sold the business and retired in 1990, Bailey had seen his company grow from a one-man, one-truck outfit to one with a fleet of trucks and other heavy equipment, employing a few dozen people and helping to re-establish a local economy and rebuild a community that had been so badly battered by an unprecedented natural disaster.

The company is still a key player on the industrial scene in Cameron Parish and still carries the name of its founder. These days, though, Shadd Savoie is the president of Roy Bailey Construction, running the business from headquarters situated between the Intracoastal Waterway and the Cameron Prairie National Wildlife Refuge at Gibbstown. The company's clientele includes the parish government, which needs a steady supply of limestone, dirt, and clay for infrastructure upkeep, and residents who are making improvements to home sites, however the company's bread and butter remains servicing oil and gas drilling locations across the vast marshes of extreme southwestern Louisiana. At any given time, it can have workers and equipment spread out at job sites across hundreds of square miles of Cameron Parish wetlands.

Entrance to Roy Bailey Construction, on the Intracoastal Waterway near the Gibbstown bridge.

Hurricane season brings with it certain rituals for Cameron Parish businesses, not unlike what residents go through on a personal level. The logistics of an actual evacuation can be more complicated, because employees involved in securing a business place and its equipment also are going to have their own families and property to protect. If there is any inherent advantage that the hurricane-prone Gulf South has over the "tornado alley" of the Midwest, the earthquake belt of California, and other cataclysmically challenged regions, it is that hurricanes come with ample advance warnings for those inclined to heed them.

In post-Audrey Cameron Parish, people heed them.

"Every year, you watch," Savoie said of tropical storm activity in the warm-weather months. Between 1990 and 2005, hurricane threats caused Savoie on three or four occasions to shut down operations and direct his employees to move the trucks and other heavy equipment out of harm's way. Better safe than sorry, they knew, but in each instance the storm had little impact on the Cameron-Creole area.

"It seemed like every time we'd bring a little less, thinking, last time it didn't come," Savoie said. "When Rita turned up, we were thinking it was going to Corpus Christi—hey, we might get some water. So we moved the equipment to higher ground, but not near high enough. We took out the vital paperwork and computers— the usual precautions. But we were thinking it's going to Corpus Christi."

As the storm began to turn toward the state line at Cameron Parish, Savoie called one of his employees and told him, "It don't look good."

He was right.

After massive Hurricane Rita skipped past Cuba in late September 2005, Cameron Parish started drawing it across the Gulf of Mexico like a magnet, its once-westward path skewing ever northward. By the time Cameron Parish residents were advised to move out, they didn't need much persuading, because the Hurricane Katrina catastrophe had heightened their sense of alarm.

Everyone had watched that disaster unfold on TV over the preceding three weeks, and most were shaken or moved by it. About 250 evacuees from New Orleans were welcomed at a shelter set up at Johnson's Bayou. Others were taken in at South Cameron Memorial Hospital, where several students from the nearby high school began volunteering, playing cards, board games, and hide-and-seek with the displaced children who otherwise had nothing to do all day but sit in a hospital room and watch television. Forty-five children from New Orleans-area families that had been flooded out were enrolled in Cameron Parish schools. Folks launched an "It Could Have Been Us Fund" to be designated for Katrina relief, mindful of their own vulnerable landscape and the capricious nature of storm trajectories.

The usual evacuation was ordered, and Hebert would estimate later that 99 percent of parish residents complied. A few holdouts stubbornly remained in locations such as Hackberry in the northern part of the parish, nominally higher than the coastal communities but still perilously close to the near-sea-level lakes and marshes of Cameron Parish's midsection. Recognizing that Rita was shaping up to be as powerful a storm as any to threaten their region in decades, most residents retreated northward, to hotels, relatives' homes, or church-based shelters in places like Kinder, Alexandria, and Shreveport or across the state line to Jackson, Little Rock, Memphis, and beyond.

Rita inundated the Cameron Parish coast with a tremendous fury, but with towns from Grand Chenier to Sabine Pass already deserted, there was no loss of life—a stark and gratifying contrast to the Audrey experience nearly a half-century earlier. The damage was so extreme, though, that as the days after Rita turned into weeks and then to months, no one was sure whether the communities of the parish would themselves survive.

When residents had evacuated on September 22 and 23, the landscape that faded from sight in their rearview mirrors was defined by the quaint homes and boats and churches typical of small Cajun towns throughout South Louisiana. After September 24, the defining characteristics of the area were storm debris and marsh grass.

Hurricane Rita didn't just damage the communities of Cameron Parish; it obliterated them. The houses and camps at Holly Beach were washed away, period. Houses throughout Cameron were destroyed. There was almost nothing left in Creole, Grand Chenier, or Johnson's Bayou.

The heavy-duty parish courthouse survived Rita as it did Audrey: "That's our rock," Hebert said. But few other buildings of any sort were left standing, and those that were—some churches, a few large homes, a handful of businesses—were just wrecked.

Among the first to view the carnage was a team of workers from Roy Bailey Construction. Before the storm hit, the company had moved much of its heavy equipment from a Creole vehicle yard to the relative safety of its rock pile near the Gibbstown bridge. On the day after Rita made landfall, Savoie assembled a crew there, and the workers set out to clear Louisiana 27—heading south from the bridge to Creole, then east all the way to Cameron.

Savoie's motivation was clear: "We needed to clear the road, so let's start clearing the road," he said. "I had the equipment sitting there. We had to get down there. This is where our livelihood is." The military could clear the highway eventually, he knew, but it would take days, at least, for the Army to move the necessary equipment into place.

"We didn't know if FEMA was going to pay us," Savoie said. "We weren't worried. We didn't even care. We just need[ed] to clear the road . . . and we started."

The highway there is set higher than the surrounding landscape, which was still covered by floodwaters, but the storm had

Cameron, Sunday, September 25, 2005.

left a thick blanket of floating turf, known as flotant, washed up from the marsh. The workers embarked on two front-end loaders to push debris off the highway, followed by an excavator for moving utility poles and other large obstructions from their path.

A full-size mobile home was blocking the highway at the little bridge on the north side of Creole. With no idea where it had come from, they moved it to the side of the highway, but only after taking snapshots of the Jesus figurine that remained upright on the mantel above the trailer's fireplace. Blue-and-white statues of the Blessed Virgin Mary were familiar lawn ornaments at homes throughout the predominantly Catholic region, as elsewhere in Cajun country; time and again, the Roy Bailey crew was amazed to see the Blessed Mother standing resolute in the floodwaters, while the homes she once adorned had disappeared. Meanwhile, as they continued toward Cameron, they struggled to keep their bearings, for virtually everything they had known along the road had been washed away.

"We'd be going down the road and we would stop, and we'd be lost," Savoie recalled. "We knew we were headed to Cameron, but we were trying to figure out where we were. I grew up here, lived my whole life here, but there were no landmarks. Finally we would notice something, and we'd say, 'OK, here's where we are.'"

Reaching the more populated parish seat, they found whole buildings and large power lines in the middle of the road.

"It looked like it was clean," Savoie said of Cameron, "because the water still covered everything. Once the water went down, then you started seeing the debris and the blocks and everything—as far as you could see. Between Creole and Cameron, it was only three or four houses we felt may be salvageable. It wasn't many. Even the houses that were intact, when you opened them, they were full of mud."

With Army and National Guard units and out-of-state law enforcement officers brought in to augment the Cameron Parish Sheriff's Office, the parish was locked down. Residents who had been sent away for a few days to avoid Rita's landfall now were kept away for weeks, which turned into months. Officials wouldn't let them back, because there was nothing for them to come back *to*. Homes, schools, churches, businesses, the library, and the hospital—all were destroyed. Eventually, residents were allowed in to

Clifton Hebert.

"look and leave," just to eyeball whatever remained of their homes, then return to wherever they were living in exile.

With no electricity, no running water, no sewer service, and no amenities of any kind in the coastal communities, authorities continued to keep the area off-limits. For many, the shock of the storm's devastation turned to impatience and frustration. Most residents wanted to return, and that meant returning to live on their own property, by whatever means necessary.

"We kept our people out for about six or eight weeks, and I tell you we were almost strung up for it," Hebert said. "We had to get all the utilities back in place—the water, the sewer, the electrical. Our people were very understanding, to some extent, but toward the end, it was, 'Let us home or else. We're ready to go.' "

A long tradition in those isolated Cameron Parish communities has been for families to pass their homes and land on to succeeding generations; it's highly unusual to ever see residential property put up for sale. That practice engendered an almost obsessive desire for storm victims to "go back home" to the very home sites where parents and grandparents had lived before them.

"I guess you could say it's bred in us to come back," Hebert explained, struggling to suppress a hearty laugh as he recalled what was instilled in him and others like him by parents and grandparents.

Cameron Parish natives know all too well the unfortunate truth about recovery from a major hurricane: it's a long slog. But they also know, either from first-hand experience or by genetic transfer, that body blows to the heart of a small community might be slow to heal, but heal they do.

"They just put on their boots and went right back to work," Hebert said. "It's been nothing but, 'When we get back to normal.' It's not *if* we get back to normal, *can* we get back to normal. It's literally *when* we get there. They've got one sight in their mind—that's the recovery."

And for many, that meant more than just getting back to their family property. Curiously, many residents were insistent about restoring the exact footprint of what Rita had washed away. And if federal regulations wouldn't allow them to rebuild their house at ground level in that floodplain, then they brought in a manufactured home, or a mobile home, or a travel trailer onto that spot. And it had to be sited *right there*.

"It's like a mindset they have: that was Grandma's house, I've got to put it back the way Grandma had it," Hebert said. "It's a funny thing, but, whew, it was a nightmare for awhile. We were trying to demolish what was left, clean up people's yards, and there's their trailer stuck right in the middle of it. You've got a whole, big area over there that's clear, but no, no, that's not *home*. I want to be *home*: home is on my house pad!"

Almost two years after Rita struck, Cameron Parish paused its recovery efforts long enough to mark the fiftieth anniversary of Hurricane Audrey. Hundreds of residents, many of them Audrey survivors by now in their fifties, sixties, seventies and even eighties, gathered in the shade of the live oaks on the courthouse lawn on June 27, 2007, for a solemn program. They sang "Eternal Father, Strong to Save," "God Lead Us Along," and "Amazing Grace." Guest speaker Robert LeBlanc, emergency preparedness director in neighboring Vermilion Parish, had taken part in search and rescue operations in Cameron Parish in Audrey's aftermath. "The sights I saw that day while rescuing people, I never want to see again," he told the crowd. "It was total destruction."

Although Hurricane Audrey is remembered every June 27 in Cameron Parish, officials suspected the fiftieth anniversary program would be the last such commemoration that many of the elderly survivors would live to see.

Across Cameron and into neighboring communities, most of the Hurricane Rita debris had been cleaned up and hauled away by that point; home sites consisted of trailers in some places, empty slabs elsewhere. Houses—real houses built from the ground up, not hauled in by truck—were still as hard to find as gas pumps and take-out food. Trosclair Road out of Oak Grove was bordered by a carpet of tranquil yellow wildflowers, but that only served as a perverse welcome mat for a succession of driveways that led to hauntingly vacant lots. In most cases, the folks that had lived there were still wrangling with insurance companies, FEMA, and the state's Road Home program to sort out what they would be allowed to rebuild there and how they would pay for it.

"There's just very little waiting on the government to come in and help, I promise you," Hebert said. "It's been straight out, 'Pay me my money for my insurance that I paid for forty years, and I'm going to build what I can, when I can, and how I can.'"

Primeaux subdivision in Cameron was wiped out by Hurricane Rita—virtually scoured clean. The rebuilding of actual, permanent houses in the neighborhood, as elsewhere in the Cameron floodplain, was complicated, though, by the federal government's insistence that they be elevated to guard against future flooding. For many, the increased cost of rebuilding at the new height requirements, instead of ground-level on their existing slabs, was more than they could afford, exceeding the reimbursements they got for their Rita losses. Yet many, if not most, of the longtime residents of the subdivision returned to their home sites, planting campers, trailers, and manufactured homes where their houses used to be.

"A lot of houses stayed for Audrey. This storm here took 'em all," said retired crane operator and Audrey survivor Wilton "W. A." LaBove.

Sitting in a folding chair alongside her husband in the shade of their camper, on the slab where their three-bedroom home once stood, Toulay LaBove was content. She nodded to her yard.

"A plate and a statue were all we found," she said. "A statue of the Virgin Mary."

THE ACADIAN
CONNECTION

When Henry Wadsworth Longfellow published "Evangeline: A Tale of Acadie" in 1847, both poem and poet became international sensations. In dactylic hexameter, Longfellow captured the imaginations of readers young and old, throughout the United States and Canada, with the stirring but fictional travails of star-crossed lovers Evangeline and Gabriel. In doing so, he immortalized the previously little-known story of the forced expulsion of the French-speaking Acadians from their homeland by the British in the mid-1700s.

Twenty-first century students don't memorize and recite the epic poem as American schoolchildren did a century earlier ("This is the forest primeval . . ."), but the image of Evangeline remains a touchstone for Acadian descendants in both Louisiana and the Maritime provinces of Canada. Statues memorializing the character highlight Acadian pilgrimage sites at both St. Martinville, Louisiana, and Grand Pré, Nova Scotia. Residents of Nova Scotia's "French Shore" along the Baie Sainte-Marie stage a dramatic presentation of the Evangeline story for tourists every summer, while in bayou country, their Cajun cousins invoke the Acadian maiden's name for everything from apartment complexes to funeral homes, from a race track to sliced bread.

Although the iconic figure is still well-remembered all these years later, Longfellow's Evangeline, while quietly dignified, was essentially a passive victim. A more appropriate role model for the Acadian people—a real-life cultural heroine—would be Madeleine LeBlanc.

Madeleine was just a child when her family was stripped of its possessions and deported from Grand Pré to Massachusetts in 1755. After England ended its Seven Years War with France in 1763, it began to allow surviving Acadians to return to Nova Scotia, but only

A visitor photographs the statue of Evangeline in front of the memorial church at the Grand Pré deportation site.

in small groups and only to the province's most inhospitable locales, not the lush farmlands of the Annapolis Valley where their families had thrived for more than a century prior to the deportation. According to an oral account cited by the Université de Moncton in New Brunswick, the LeBlancs traveled back to Nova Scotia in 1772 to make a new home on the rugged coastline of Baie Sainte-Marie.

When they landed near what would become La Pointe de l'Eglise (Church Point), "they were devastated to see the difficult conditions," related Barry Jean Ancelet, professor of Francophone studies and director of the Center for Acadian and Creole Folklore at the University of Louisiana at Lafayette. "It would be impossible for them to farm as they had in their previous lands. The entire family sat around the boat and wept."

After a while, Madeleine, in her mid-twenties, picked up an axe, made her way out of the vessel, and set off to chop down a tree.

"Someone asked her what she was doing," Ancelet said. "She answered, 'J'avons pleuré assez; c'est l'heure de couper du bois.' (We have wept enough; it's time to cut some wood.) And the family went on to firmly establish itself by working hard and figuring out a way to survive in the conditions they were in. They and their subsequent neighbors found themselves in the woods and along the rocky coast, so they became lumbermen and fishermen and they thrived."

Among the Acadian descendants in Louisiana's Cajun country, Leola Terrebonne Trahan of Delcambre is a modern-day Madeleine LeBlanc.

Trahan worked twenty-five years for the Vermilion Parish Clerk of Court's Office. Three months after she retired, Hurricane Rita took almost everything she had. Her home was destroyed, its contents ruined. Her furniture, most of her clothes, the irreplaceable photographs of her parents—all lost.

Three months later she was back on Judy Street, in a FEMA trailer. Thirty feet long and eight feet wide, the cramped government-issued camper would be her home for the next fourteen months. She made do there without complaint, although she did allow herself one creature comfort.

"The mattress that comes in there, I never slept on a tomb but I would think that's what it would be like, it's so hard," Trahan said. "So I bought me a new mattress."

Her neighborhood, her town, and her life were all in shambles. Not sickly by nature, Trahan endured a variety of maladies—a chronic cough, congestion, sinus infections, and nose bleeds—which she suspected were triggered by mold that erupted in the trailer after rain started leaking into it. Sometimes she would break down crying, overwhelmed by her lonely situation in the middle of a post-apocalyptic landscape that no longer resembled the Delcambre she had known since childhood when her shrimper-father moved the family there from Bayou Lafourche.

Then, at some point, she willed herself beyond the despair. After navigating the shock, the grief, and the self-pity, Trahan decided she had cried enough and the time had come to make the best of things and move on. She started picking crabs and selling the meat to bring in a few dollars. A long-time notary public, she found there still was money to be made notarizing records for local residents, but she also recognized that everyone was struggling in the town where only a handful of homes had not flooded, so she only charged half-price for her services. Depleting her modest nest egg, she bought a mobile home to replace the home she lost.

"When you think about it, you work all your life and then it's gone," she said. "But then you say OK, and you get that out of your system. Life goes on. I'm not going to sit here and continue feeling sorry for myself. No, no. We didn't do that, and I think that's what made us strong."

There's no denying that the people of lower Acadiana showed great resolve in rebuilding their communities from two major hurricanes, Rita in 2005 and Ike in 2008. Doing so away from the spotlight that stayed trained on post-Katrina New Orleans, they demonstrated that perseverance is in the Cajuns' nature, and it served them well in those times of crisis.

Anyone familiar with the history of the Cajun people can aver that they come by that character trait honestly.

Hurricane Rita was unusual, since the storm took a destructive swipe across the entire length of the coastal Cajun region of Louisiana instead of targeting one locale. For such a powerful storm to do so in September 2005 was a coincidence of cosmic proportions. It was 250 years earlier, to the very month, that dramatic and tragic events were set in motion half a continent away—events that would lay the groundwork for the very creation and development of Louisiana's Cajun culture.

In an era when European powers were constantly jockeying for advantage in the expansion of their respective empires in the New World, control of what would become the Canadian province of Nova Scotia on the Atlantic coast volleyed from France to Great Britain ten times between 1605 and 1713. The first Acadian settlers had arrived from France in the early 1600s and built on North America's first frontier a successful agrarian society, working together on communal projects such as their uncommon system of diking the tidal marshes to convert them to bountiful farmland. In the land they called Acadie, they lived in harmony with the native Mi'kmaqs and traded with colonists who populated areas to the south and west.

They saw themselves as neutrals in the global tug-of-war between the two superpowers of the day, but because they spoke French and clung to their Catholic faith, the British eyed them with great suspicion. In the run-up to the Seven Years' War, British leaders in Nova Scotia demanded that Acadians take an oath of allegiance to the crown. Most refused, primarily on religious grounds, and the colonial British government responded by expelling them from

At the Acadian Memorial in St. Martinville, those exiled from Acadie in 1755 are memorialized.

the province, beginning in September 1755. They were stripped of their possessions, their homes and farms were given to loyal British subjects brought in from other colonies, and they were dispatched to inhospitable port cities all along the Eastern Seaboard, or prison in England, or unsuccessful repatriation back in France.

Many exiled Acadians were separated from their families, many died from inhumane conditions to which they were subjected, and many who survived the ordeal were shunned in their new surroundings. The events became known as *Le Grand Derangement*— the great upheaval. Historians from Louisiana to New England have described the Acadian expulsion as the first instance of ethnic cleansing in North America.

Tragic though it was, the dispersal made possible the eventual migration some years later of small groups of surviving Acadians to South Louisiana. Although they found it hot, humid, and maddeningly mosquito-infested compared to the homeland they had known prior to the deportation, the Acadians were welcomed there and adapted to their new surroundings, as did a multitude of other ethnic groups who arrived on the Louisiana frontier in the 1700s and 1800s. As time passed on the bayous and prairies of South Louisiana, that curious mix of peoples percolated in isolation from the rest of American society, giving rise to a Cajun culture that by the late twentieth century had become a Louisiana trademark. With its signature cuisine, music, language, and zest for living, it—like the international city that is New Orleans—set Louisiana apart from the rest of the Bible Belt South.

For many contemporary Cajuns, links to the past are well-documented in church and civil records and sustained informally over successive generations of distinctive folk practices. Moreover, those connections provide them with a collective identity that is celebrated today, but only after enduring more attempted purges in the twentieth century. Older residents across Acadiana can attest to being harshly punished during long-ago school days for speaking their native tongue and being ostracized by English-speaking classmates and teachers. For many years, as modernization and "Americanization" infiltrated once-isolated communities, Louisiana's French-speaking people found it necessary to circle the wagons, suppress outward demonstrations of their customs in order to get along, and keep their peculiar ways to themselves.

"When the British exiled us from Nova Scotia beginning in 1755, it was with the expressed intent of dispersing us among the British colonies so we might be absorbed and acculturated. This did not happen," Ancelet said at a symposium on the Acadian diaspora, held in New Orleans four months after Hurricane Ike. Ancelet continued:

> Instead of eliminating the Acadian identity, the exile further galvanized it.
>
> Those Acadians who arrived in Louisiana between 1765 and 1788 were expected to dissolve into French Creole society. This did not happen. We preserved our cultural and social specificity well after the French and Spanish periods.
>
> Under pressure from the fierce nationalism that accompanied World War II, we were expected to melt into the American melting pot. This did not happen. Cajuns found ways to negotiate the mainstream and continued to celebrate our traditions and language.

Thus, when the Cajun communities across the coastal region of South Louisiana—places like Dulac, New Iberia, Delcambre,

Kaplan, Grand Chenier, and Cameron—were devastated by Hurricane Rita, and again three years later by Hurricane Ike, those were more than just natural disasters. In the bigger picture, those storms posed merely the latest assaults on a people and a way of life whose tribulations date back more than two-and-a-half centuries.

Tangible remnants of that history were placed at risk as well when Hurricane Rita took aim at the Acadian Museum in downtown Erath.

The museum, housed in a sturdy old bank building, had become a depository for documents, photographs, audio and video recordings, articles of clothing, political memorabilia, and other mementos of the life and times of the area's Cajun people. The driving force behind the museum was Warren Perrin, a native of nearby Henry and the longtime president of the Council for the Development of French in Louisiana. Perrin was a partner in a Lafayette law firm, but he kept office hours at the museum one or two days a week, invariably speaking French with his older clients and drop-in visitors.

In 1989 he had launched a quixotic legal battle aimed at extracting an apology for the deportation of the Acadians from Great Britain's Queen Elizabeth II. Perrin was a tireless champion of Cajun culture, and no one who knew him was at all surprised to see him take on such a challenge. Most viewed the petition as a symbolic gesture, but Perrin was utterly serious about his endeavor. Sure enough, after fourteen years and countless pleadings along a meticulously navigated path through the British legal system, Her Majesty apologized.

Technically, it was more of a don't-call-this-an-apology apology. The queen's representative in Canada, Governor General Adrienne Clarkson, signed a royal proclamation acknowledging the expulsion by British forces and its "tragic consequences, including the deaths of many thousands of Acadians—from disease, in shipwrecks, in their places of refuge and in prison camps." The proclamation also designated July 28 as a date for commemorating the Acadians every year. It was on July 28, 1755, that a British military officer signed the deportation order.

The crown stopped short of saying, "We're sorry," but Perrin felt vindicated.

"It's never too late to right a wrong. I feel a lot of relief for Acadi-

Warren Perrin discusses the history of Louisiana's Cajun people with a group of tourists at the Acadian Museum in Erath.

Perrin stands with the proclamation he obtained from Queen Elizabeth II, apologizing for the Acadian deportation 250 years earlier, at the Acadian Museum.

an people everywhere," he said when the proclamation was issued. "At the time I drew up the petition, I didn't know what I was asking for, a cross maybe, or a pile of bricks somewhere. To have a day to commemorate what happened to the Acadians, this is better."

Perrin's copy of the royal proclamation would become one of the most treasured pieces housed in the Acadian Museum. Also among the most valued and historically significant items there was a collection of artifacts from a pre-deportation Acadian settlement, donated by Nova Scotia archaeologist and historian Sara Beanlands shortly before Hurricane Rita struck.

The summer after the queen issued her apology, thousands of Cajuns from Louisiana and others of Acadian extraction from across the United States, Canada, and elsewhere had gathered in Nova Scotia for the third Congrès Mondial Acadien, or World Acadian Congress. The international gathering of Acadians, inaugurated in 1994, was held every five years in a different location; the 1999 event took place in Louisiana. More than two weeks in length, the Congrès typically includes major events such as concerts, symposiums, and gala opening and closing ceremonies, but its most popular aspects are dozens of family reunions where hundreds, and in some cases thousands, of descendants of the pre-deportation Acadian families gather to celebrate their shared history.

The Thibodeau family reunion in 2004 was held at the deportation memorial park in Grand Pré. Beanlands caused a sensation there, and at her family's farm about twenty miles away the next day, by leading Thibodeau descendants on a journey through space and time to walk the grounds where some of their Acadian ancestors had lived prior to the deportation—a site that had been in her family's care ever since.

Willow Brook Farm has always been an idyllic spot for her. Growing up a city girl in the provincial capital of Halifax, she spent her summers there, cavorting with her cousins, clambering over abandoned farm equipment, and hiding in lopsided old barns. For Beanlands, there was a comforting feeling about the farm—the hills and pastures, the family togetherness, the farmhouses and barns, even the cows. This was the Shaw family farm, run by her uncles, Allen and David Shaw, and before them by her grandfather, Anthony Shaw.

As she got older, she developed a keen interest in history, and she began to take a more scholarly interest in her family's background. Her research revealed that Arnold Shaw had been a successful farmer in Little Compton, Rhode Island, until he was recruited by the crown to relocate in Nova Scotia in 1761. Like other hand-picked settlers, he received a land grant for one of the area's best farming sites.

The farm has been in the Shaw family since then, a rare example of a property remaining in the possession of direct descendants of one of Nova Scotia's earliest "New England planters." Uncles Allen and David were the seventh generation of Shaws to farm the property. Shaws who preceded them there are buried in a family cemetery in a grove of trees on the farm.

Beanlands' research had been easy, thanks to a bountiful paper trail recording every significant development along the way, back to Arnold Shaw's acquisition of the farm. Whatever came before that didn't seem to matter. Then one night in March of 2003, she and her parents were back on the farm for a family dinner. Over the course of the evening's conversation, her Uncle Allen mentioned that he had gotten a recent phone call—out of the blue, after eighteen years—from Dick Thibodeau.

Beanlands' ears perked up. Who was that? What did he want?

Allen Shaw casually told his niece what older family members had long known.

Dick Thibodeau was a retiree from New England who had turned up at the farm one day back in 1985, clutching a copy of a faded old map and looking for a spot where he believed his Acadian ancestors had lived prior to the deportation. Allen had recognized the features on the map, including a distinctive bend in the St. Croix River, and showed him around the farm, helping him locate the places that corresponded to the five dots on his map.

The map was dated 1756. The dots were labeled "Thibodeau Village."

The man was grateful beyond words. After he'd seen enough, he went home to Massachusetts, Allen went back to the work, and that was that.

Beanlands had been thirteen years old when Dick Thibodeau had come to Willow Brook Farm. Had she even known about his quest at the time, it would have meant nothing to her. But now, as a family historian and a college history major, she found it difficult to take this all in: An Acadian village? On our farm? When? How? Who?

Sara Beanlands and her uncle Allen Shaw walk across the Shaw family farm at sunrise.

Beanlands was enthralled by the story her uncle related. No one had ever said anything to her about Acadians having lived there before the Shaws.

"You have to understand something about this area: It's English to the core," she said later. "You wouldn't find a French-speaking person there to save your life."

As she thought about it, though, she realized there were clues, scattered all over Willow Brook Farm. There were the French coins that got plowed up from time to time. The neat patch of flowers that bloomed in the middle of a pasture every year, where a home must have been long ago. The trail through the farm, down to the river, that locals call the Old French Road. The spot everyone knows as French Orchard Hill. Even the name Willow Brook Farm harkened back to the willow trees that were brought from France by the Acadians. Beanlands had never put it together before, but her uncles knew.

"Nobody knows his land like a farmer," she said. "These stories were passed on from father to son. This was all just not very important to my family at the time. They're farmers. They're good people, and they're good at what they do, but there was just not a good understanding of the history of Nova Scotia or the Acadians."

Now the niece studying archaeology and pursuing a master's degree in history understood.

"I am loyal to my family," she said. "I want their lives to continue as they used to. But I feel compelled to record the Acadian history that is here."

The first thing Beanlands did was send a letter to Dick Thibodeau, to get his story first-hand. He called her immediately, and a long-distance partnership was born. Then she immersed herself in the Thibodeau family history he provided. She arranged for university archaeologists to excavate for artifacts on the suspected site of one of the Acadian dwellings. The dig on the hill overlooking the river revealed a wide array of utensils, smoking pipes, and other household items that dated the homesite to 1749.

Research concluded that Thibodeau Village was founded by Pierre Thibodeau in 1690. Pierre, born in 1670, was the oldest son of Pierre Thibodeau and Jeanne Terriot, who came from France to begin the Thibodeau family line in Acadie. Son Pierre and his wife, Anne Bourg, had twelve children, all at Thibodeau Village. Among them was a son named Alexis, who was separated from his family and shipped off to Philadelphia in the deportation of 1755, 230 years before his descendant, Dick Thibodeau, would walk in his footsteps on Willow Brook Farm.

As the 2004 Congrès approached, Thibodeau and Beanlands reported what they had uncovered to the Thibodeau reunion committee. They were invited to share their findings at the family reunion in Grand Pré.

That Sunday morning, Thibodeau went first, recounting his circuitous search for his roots. Then he alluded to some exciting discoveries about the family's ancestors, and he introduced Beanlands to pick up the story.

"I wasn't sure what was going to happen," she said later. "I had some really good information, but I had no idea how I would be received. I hoped for the best, but I thought people might be a bit uncomfortable, like, 'Why are you here when you're the people who took our lands?' I was hoping one or two people might come up when I was done and say thanks, but I was prepared for some negative reaction, too."

She looked out at the 325 people who had crowded into the huge white tent near the park's famous statue of Evangeline, took a deep

Sara Beanlands leads members of the extended Thibodeau(x) family on a tour of her family's farm during the 2004 Congrès Mondial Acadien.

Sara Beanlands, her grandmother, Beulah Shaw, and her mother, Hope Beanlands are all smiles as they host members of the extended Thibodeau(x) family on the Shaw family farm during the 2004 Congrès Mondial Acadien.

breath, and began: "This is a story about a landscape shared by two families with a very unique connection to the land and, by extension, to each other for more than three hundred years . . ."

Beanlands had never made a PowerPoint presentation before, and it showed. She was nervous. She got mixed up. She couldn't get the equipment to work properly. She kept bumping into the microphone. At one point she called her father up to the stage to try to help her get through it. She muddled on, concluding with a modest invitation for interested family members to drive out to Poplar Grove the next morning and visit the farm. Unable to sense what kind of reaction she was getting from the stoic audience, she feared the worst.

She need not have worried. People started lining up to talk to her before the applause had even died down. Some pressed her for more information, but all of them wanted to thank her for taking such a gratifying and unexpected interest in their family. The receiving line took half an hour to play out. There was not a single negative comment.

The next morning, Beanlands stationed herself at the Windsor welcome center to meet folks who wanted to see the Thibodeau Village site for themselves. To her shock, ten vehicles turned into twenty, then thirty, then forty. She gave it a few more minutes, then called her Aunt Joanne with a head's-up: "I'm coming with the Thibodeaus."

"What are you talking about?" Joanne Shaw replied. "They're already here!"

Thirty to forty people from the reunion had bypassed the rendezvous point indicated on the map Sara had distributed and proceeded directly to the farm. Once Beanlands arrived with her entourage, what transpired that morning at Willow Brook Farm was magical. There were perhaps 150 descendants of a long-ago Acadian family gathered in that spot, and four generations of the Shaw family went out of their way to make them feel welcome, appreciated, and at home.

People marveled at Joanne and David Shaw's Acadian-looking house, rotating in and out of the basement where they took snapshots of the support timbers and the two-foot-thick stone foundation wall. They chattered and laughed and got to know their hosts and hostesses. It became an unscheduled extension of the family reunion.

More than one hundred of the visitors followed Beanlands

down the Old French Road for a tour of the farm's Acadian sites. Most of them made it up the hill to Pierre and Anne's homesite, with its stunning view of the environs including that bend in the St. Croix River. Many lingered there, taking more pictures, admiring the scenery, soaking it all in, and connecting with the past.

"This is the site," said Beanlands, who would go on to focus her graduate research on Nova Scotia's Acadian history. "We're never going to let it be forgotten again."

Warren Perrin was in Nova Scotia for the Congrès. Not knowing what was going to transpire at the Thibodeau reunion, though, he had no reason to attend it, since his predominant Acadian family line was the Broussards. Consequently, he missed out on the transcendent gathering at the Shaw farm. Word of the event soon spread up and down the Acadian pipelines of Atlantic Canada and South Louisiana, though, and it wasn't long before he met up with Beanlands. They took an immediate interest in each other's efforts to explore and sustain Acadian history from their very distinct perspectives.

Five months later, Hurricane Rita tore through South Louisiana. As the storm made landfall, Perrin and his wife, Mary, stayed at their home in Lafayette. Rita's strong winds toppled two trees onto their house, causing considerable roof damage, but their concern was for the Acadian Museum. The next day, the Perrins made their way to Erath in the hope of salvaging some of the museum's most valuable holdings. The flooded town was locked down by the military, but after some measured negotiations, they were given one hour and one truck to rescue what they could.

The losses were staggering. The hurricane had dumped two and a half feet of water into the building, and every time a boat would pass en route to or from the town's rescue staging area nearby, more water sloshed inside. File cabinets filled with irreplaceable historical documents, photographs, genealogical charts, and vintage newspaper clippings were inundated, their contents trashed. The lending library of books, audio recordings, and videotapes was destroyed. Mold and mildew were poised to ruin everything else.

They retrieved the queen's proclamation, which was unscathed; a round miniature painting which was believed to be the first depiction of Longfellow's Evangeline; some pottery from the home of the French explorer Samuel de Champlain; an antique christening gown; and the museum's collection of hand-woven textiles. Fearing that the textiles would get mildewed if left unattended, Mary Perrin hand-washed each piece as a museum conservation expert recommended.

"The old Acadian home-spun and hand-woven blankets were made to withstand almost anything except mildew, but each one must have weighed twenty pounds when wet," she said. "It was hard work!"

Beanlands made her first visit to Louisiana the next spring, and she brought with her a few gifts from the family farm. She donated to the Erath museum several pre-deportation artifacts that had been unearthed from Thibodeau Village: coins, musket balls, a pipe stem, a button, and a shoe buckle. Perrin gave them a place of honor alongside the queen's proclamation in the museum's Acadian Room.

During her trip, Beanlands also spent three days in the New Orleans area and got a first-hand look at the Katrina devastation there. Her lasting memory, though, was from Cajun country, where the wounds from a massive hurricane were already starting to heal.

"I guess what I was struck most by was the sense of community. There was a silent strength about Erath, which I attributed to people who could rely on each other," Beanlands observed later. "Perhaps that strength comes from a people who have historically understood the value and importance of community to survival."

WHERE TRADITION IS CULTIVATED

For Ernest Girouard, farming wasn't just a science—it was an art form. Rice was his medium, the fertile soil of Vermilion Parish his canvas.

He had a genuine talent for planting a seed, producing a crop, and nurturing that bounty until harvest, and he put that talent into practice all day, every day, throughout the growing season, year after year. Taking a plot of land and making it more productive was more than an occupation for him. On a deep, metaphysical level, Girouard considered it a calling.

"Being out there in the fields every day is something that I love to do," he said.

Girouard came by his passion honestly. His Acadian ancestors had been farmers, dating all the way back to Francois Girouard, who arrived in Port Royal (now known as Annapolis Royal in Nova Scotia) from Varennes, France, in 1646, married Jeanne Aucoin the next year, and founded the family line in the New World. The Girouards settled across from the fort at Port Royal on the north bank of the Dauphin River (now the Annapolis River), where they raised crops, cattle, and sheep, and reared five children.

A century later, when the British began deporting Acadians from Nova Scotia, the extended Girouard family was split three ways. Some fled deep into the forests of neighboring New Brunswick and avoided the deportation. Others took a more circuitous route and ended up resettling in Quebec. Ernest Girouard's direct ancestors were imprisoned for a time in England, then were repatriated to France and eventually made their way to Louisiana, where land grants issued by the Spanish governor in New Orleans provided them with a fresh start.

Many members of those Canadian families drifted down to New England and New Jersey in the mid-1800s in pursuit of factory jobs as the textile mills there expanded. The Louisiana Girouards, meanwhile, worked the land of the southwestern Louisiana prairie along with other Acadians and settlers of other ethnic backgrounds. And beginning in the final quarter of the nineteenth century, what had been a fledgling rice industry began to flourish in places like Vermilion Parish and its neighbor to the north, Acadia Parish. In those and surrounding areas where land was available and well-suited for rice production, Louisiana's rice farming activity became firmly established at a time when sugar prices dropped precipitously and the boll weevil ravaged the state's cotton crops.

Amid those developments, Ernest's grandfather and great-uncle homesteaded the family estate near Kaplan. His father made a career as a rice farmer, and Ernest helped to carry that farming tradition all the way into the twenty-first century. The first person in his family to go to college, he emerged from Louisiana State University in 1968 with a doctorate in dairy science. He initially took a job as an animal nutritionist for a feed company in Illinois, but the next year he returned to Vermilion Parish, where he and his brothers joined their father on the family rice farm. After thirty-two years as a partner in the family farm, Girouard began a new rice farming operation several miles below Kaplan in the remote southern part of the parish, mostly on the southern side of the Intracoastal Waterway.

"It was a tremendous experience to be able to go down there and spend an entire day on land that was almost isolated," he said. "There weren't many homes down there, and it was ideal for the production of rice."

The area had an abundant water supply, unlike other farm sites in many parts of Acadiana where farmers could only irrigate their fields by drilling wells and pumping water from far underground. Girouard developed a system for routing the available water into his rice ponds and a method for recycling that water to get the most use out of it. He used a laser to level the ground with extreme accuracy, ensuring that the rice stalks in each leveed plot would be covered with the optimum amount of water—just two to three inches—during the growing season.

He got tremendous satisfaction from the improvements he made

Ernest Girouard in a Crowley rice field.

to the land and the crops he produced as a result. What's more, there was something about that farming community in the lower part of Vermilion Parish that he found particularly gratifying.

"I really appreciate the landowners and the local people down there on how they were really self-sufficient," he said. "They helped each other. They were very independent. They had strong, strong family values, and their children were all involved in the ways that their parents made a living."

As he practiced his farming artistry, Girouard—by then in his sixties, and well into his fourth decade as a farmer—was happy to be among these salt-of-the-earth people. Then Hurricane Rita struck in the fall of 2005. All of a sudden, his future as a farmer was threatened by, of all things, the salt of the earth.

Because rice fields are bordered by levees on all sides, the storm surge that overtopped those levees sat in those fields with no way to drain out after the floodwaters subsided elsewhere. Consequently, the ground underlying the flooded fields absorbed prohibitive amounts of salt from the invasive seawater.

"When the storm surge came in and flooded the land that I farm down there, the surge came up the Vermilion River in the east and then moved west to the land that I farmed," Girouard recalled. "It reached the levees and went over the levees, and then we had water on the inside that our fields trapped. In order to get that saltwater off the land, we had to breach the levees ourselves with our equipment. The land had saltwater on it for thirty or more days longer than on the outside."

Farmers measure salt content in the soil by parts per million. Rice farmers need to keep the salt content down to 750 ppm for unrestricted growth and to a maximum of 1,000 ppm to get any production out of the soil at all. After the long-term inundation brought on by Rita's storm surge, the soil on Girouard's farm measured 8,000 ppm in some places, 5,000 ppm in others.

He would not be able to grow rice there—for a long time.

He wasn't alone. Sugarcane farmers, cattle ranchers, and other rice farmers throughout southern Vermilion Parish were hit hard—harder than ever—by the flooding. Collectively, their struggles with Rita and its aftermath proved to be the biggest challenge faced by the parish's vital agricultural community in their lifetime, exceeding even the benchmark impact of Hurricane Audrey almost fifty years earlier.

When it came to hurricanes, farmers and others in the region tended to be most fearful of wind damage, but there was very little of that from Rita. It was the storm surge that destroyed houses and barns and fences, and what the flooding didn't knock down, the salt in the floodwaters ruined. All types of farming equipment—tractors, pumps, generators, and alternators—succumbed to the effects of the storm surge.

"No one anticipated the saltwater would corrode so quickly," said Howard Cormier, longtime agricultural agent with the Louisiana AgCenter extension service in Abbeville. "Wire fence posts corroded where the wire was attached. The fence posts just turned to rust very quickly. The image of seawater being that corrosive, most people don't think of it, but that's what happened here. While the farmers were trying to deal with getting debris off their land, those who were flooded, when they'd go to start a pump or any piece of equipment, they found the wiring harness was eaten away—or the starter, the alternator, the battery, all that."

Sustaining a ranching tradition that dated back to the 1700s,

Floating marsh and a farm building washed onto a road by Hurricane Rita.

Vermilion Parish residents lead cattle through the floodwaters to higher ground after Hurricane Rita's storm surge hit.

Vermilion Parish was the largest cattle producer in the state, and the many cattlemen in the flood zone had their own concerns. The water stayed up for days, and cattle that didn't drown had scattered or were stranded. Feed, hay, and water for cattle and horses became instant priorities; not only did stockpiles of hay get washed away, but salt in the floodwaters killed the grass in pastures throughout the prairie, up to twenty miles inland from the coast, so those cattle that did survive the flood were left with no place to graze once the water receded.

"Our immediate challenge was getting feed from somewhere, feed and hay, brought in and then dispersed to the people that needed it, and developing a system to disperse it fairly," Cormier said. "We also took charge of providing drinking water for horses that were used to work cattle out in the marsh, because the horses would work all day and had nothing to drink. Someone on our staff got a 1,500-gallon tank and they would fill up at the city hall here. As people needed, they would call, and they would go and water twenty, thirty horses at one point in time. Those horses could make it the rest of the day when they had a good drink of water."

Long before federal emergency personnel ever arrived to begin their formal, deliberate, and often ineffective relief efforts, the initial response in the immediate aftermath of Hurricane Rita came from neighbors helping neighbors—sharing provisions, sorting through and clearing away debris, doing little things to help each other get from the chaos of impact to the first incremental steps toward recovery. Nowhere in Acadiana was that commitment more aptly demonstrated than up and down the flooded country roads of Vermilion Parish as men and boys fanned out on horseback, in boats and pickup trucks, and on four-wheelers to help each other round up livestock and secure their property.

Working in the floodwaters, though, presented problems that they had never experienced and certainly didn't anticipate.

With the flat landscape of the farming and cattle-grazing region covered in water, men on horseback couldn't distinguish the sides of the roads. Horses often dropped off suddenly into flooded ditches, leaving both horse and cowboy to swim for higher ground.

Many of the men used airboats to reach cattle that had been stranded on the ridges known as "islands" in more remote locations in lower Vermilion and Cameron parishes.

"When the airboat would drive up to the island, some of the cows would try to get on the airboat," Cormier said. "The cows had been starving and drinking saltwater for several days, so they were mentally unbalanced to say the least. They were very aggressive. Some of them had to be shot because they wouldn't stop attacking the cowboy, whether they were on a boat or a four-wheeler or whatever."

Then there was the bull, mired in the water, that every once in a while would start bellowing and thrashing about wildly.

"When they investigated more closely, there was a ten or twelve foot alligator that was behind him and that would take a bite out of him and rip it off every once in a while," Cormier said.

Landowners had to contend with more than just water when the hurricane flooded the region. When the storm surge came through, it swept along dense round hay bales, logs, flotant that had carpeted miles of coastal marshes, and wreckage from homes and camps it damaged or destroyed. As the surge washed across the countryside, the debris that it contained knocked down fences throughout the region. In the coming days, when cattlemen set out into the floodwaters on horseback to round up their cattle, many of their horses got caught in the barbed wire and other fencing that was submerged in the murky water.

And they weren't the only ones confronting this problematic situation. The challenge was even more daunting for sugarcane farmers.

Since Jesuit missionaries brought it to South Louisiana in 1751, sugarcane thrived in the region's temperate climate. In modern times, unstable sugar prices were offset by improved farming techniques that increased the sugar content in cane stalks and more efficient processing practices at mills and refineries, so by the turn of the twenty-first century the annual crop remained profitable, contributing two billion dollars to the state's economy.

Sugarcane season coincides with hurricane season, though, and Rita struck at a time when Louisiana sugarcane farmers had a fully mature crop in the fields—towering leafy stalks just days away from the October harvest. Wind is always the primary concern for sugarcane farmers during a hurricane, but it was the water that did them in this time.

Sugarcane is cultivated in long rows on massive open fields, unlike the smaller flooded ponds used for growing rice, so the harmful saltwater in Rita's storm surge receded within a few days. Although there was some impact from salt that seeped into the cane fields, the real problem resulted from the other things that washed in with the surge and got left behind when the floodwaters receded.

Cane farmers in the flood zones of Vermilion, Iberia, and St. Mary parishes checked on their farms after the storm passed and were aghast to find them blanketed in debris up to three feet thick. And it wasn't just the marsh grass that had been carried in from the coastal areas to the south—it was everything else in the storm surge's path. Mixed in with the flotant from the marshes to the south were bridge timbers, utility poles, and the remains of destroyed houses, camps, garages, and sheds—not just fragments of roofs and walls, but the furniture and appliances and everything else the buildings had contained.

As if the sheer volume of debris wasn't enough of an obstacle for farmers, plentiful among the mess were propane tanks from the seafood boiling set-ups that were standard equipment for homes and camps throughout South Louisiana. They proved to be particular dangers. If a farm worker operating a $250,000 piece of machinery struck an ice chest hidden in the marsh grass, that didn't cause much of a problem. If that same worker on that same piece of equipment hit a propane tank, though, an explosion could destroy or seriously damage the machine and injure the man who was driving it.

And farmers *had* to clear those fields in order to harvest what remained of that year's crop and prepare for the next one.

Errol Rodrigues farmed sugarcane in the area around Erath, maintaining a family tradition that started with his great-grandfather. Most of his 3,600 acres of cane fields were south of Louisiana 14, about six miles from the northern shore of Vermilion Bay, which opens into the Gulf of Mexico. Rita's impact on his acreage was typical for farms in the region: the land was covered by about six feet of water, and once that water drained off, it left the soil excessively salty—and coated with debris.

"There's a lot of camps that sit on the coast of Vermilion Bay— there *were* a lot of camps, let me put it that way," Rodrigues said. "All those camps broke up with the storm. The barrier that stopped all these camps and all the marsh floating was the cane fields. All those houses along there, gas cans, propane bottles, anything that wasn't tied down just floated, and the sugarcane caught it. That marsh grass just picked up all that stuff up as it floated. The flotant was three foot thick and it gathered together and landed there."

Workers scoured the fields at harvest time, removing gas tanks, refrigerators, and other obstacles they could find. After the cane was cut, the fields were burned—a common practice to prepare them for the next planting, but even more necessary that year to get rid of Rita's flotsam.

"All those fields were hand-cleaned," Rodrigues said. "As we harvested, we'd burn them. When we burned the fields, you'd hear the gas tanks [explode]. That's why we had to hand-pick the fields, because of the propane tanks. People would catch them in harvesters. One of my neighbors caught a personal oxygen tank, for breathing. It exploded. One piece went about six to seven hundred feet ahead of him, and the other piece went through the floor of the harvester. Nobody got hurt. We had men on tractors with front-end loaders. A man would ride in the bucket of the front-end loader, and one driving, and they'd follow the harvesters around the field, basically cleaning the next row that the harvester would cut."

It was a long harvest.

Although Rita's impact posed different challenges from rice pond to cow pasture to cane field, the farmers and cattlemen, their families, and their neighbors met them with a collective determination—a shared self-reliance that buttressed recovery throughout Cajun country. For members of the farming community especially, doing for themselves and for each other, individually and with the help of the local extension service, meant things got done. The work might not have been letter-perfect, but it was expedient and effective and then folks moved on to the next thing that needed doing.

"Everyone went to work right away, from the time they got back. One thing we're proud of is that we didn't have finger-pointing. We didn't blame the mayor for not doing anything, the fire chief for not doing anything. We didn't blame the governor," Cormier said.

"If there was a beauty to this whole deal, it's a story of a people that pulled together, unselfishly, for the good of the group. There was no looting that I was aware of. Neighbors took care of watching out for each other. We didn't spend much time wringing our hands and feeling sorry for ourselves and saying, 'Who's gonna help us?'

It was good to go through the deal in this area rather than where you had to fear for your life and belongings in some other areas."

As much as that camaraderie helped local farmers to secure their homes and property, though, no amount of neighborly concern was going to make the rice start growing again. High salt content in the soil prevented seeds from germinating at all, or stunted what little growth did occur. The dire situation would linger through 2006 and into 2007 as the area suffered from a lack of rainfall which would have flushed the salt from the ground.

Throughout history, the world over, those who work the fields have had to endure the vagaries of the elements, so difficulties brought on by the weather were, in a general sense, nothing new for the farmers of Vermilion Parish. The impact of salt contamination from Hurricane Rita's floodwaters was particularly frustrating, though, because that had never happened to them before—and no one seemed to know how to fix it, besides waiting for the soil to fix itself.

Farmers' grave concerns about salinity levels in the ground even confounded the experts at the extension service in Abbeville. Members of the AgCenter staff distributed hundreds of copies of a scant, thirty-year-old pamphlet, "A General Guide For Using Salt Water On Rice"; it wasn't much, but it was all they had.

At his eight hundred acre rice farm in the southern part of the parish, Girouard set out to bring the land back into production, but his efforts were hampered by the dearth of fresh surface water that he had come to rely on to irrigate his fields. The hurricane had destroyed locks that had provided reservoirs of fresh water throughout the Mermentau River basin, and saltwater from the Gulf continued to infiltrate the area's waterways due to persistent southerly winds. Needing rainfall to wash the salt out of the soil, he and neighboring farmers were faced with an uncharacteristically dry season. By late spring 2006, the ground in his ponds was still bare. Even weeds were struggling to grow in the salty soil.

It was a difficult year for everyone in the wide open spaces of Vermilion Parish. Rita cost the parish 25 percent of its cattle production: of its 40,000 head of cattle, 5,000 succumbed to the flooding and another 5,000 had to be sold off by ranchers whose grazing lands were ruined and who couldn't provide feed for their herds. Sugarcane farmers whose fields were blanketed in marsh debris

Floodwater from Hurricane Rita's storm surge rushes through a Vermilion Parish farm.

Floating marsh known as flotant was pushed miles inland by Hurricane Rita's storm surge and dumped into this Vermilion Parish sugarcane field.

lost two growing seasons. And the salt in the rice ponds wasn't going anywhere, anytime soon.

"One of the hardest things for me was to see that productive land that wouldn't even grow but a few weeds over an entire period of a year, when in the past it had produced a very productive crop," Girouard said. "And then you didn't know whether it was going to ever be able to come back into production. Some of that land had so much saltwater that it actually started to lose its structure. Whenever a soil loses its structure, it becomes unproductive and it takes years and years to bring it back."

And many farmers couldn't wait that long. The adverse conditions, and the frustrations they engendered, prodded many in the region to give up farming.

At one time, Vermilion Parish boasted about seven hundred rice farmers. By two years after Hurricane Rita, those ranks had dwindled to about 250. And the prospects for sustaining that farming tradition into future generations appeared tenuous.

"We do have some young rice farmers, but a number of them have other jobs and their wives also work," Cormier said. "They are hanging on sometimes because their family wants them to hang on, but it's getting real tough."

Those who gave up on farming had little trouble finding work in the post-hurricane economy of South Louisiana.

"They went to work in the oil industry and the construction industry, and they started at top dollar because they knew the work ethic and the family values were carried over to their jobs," Girouard said. "Most of them became very, very successful. Most of them that I know are doing much, much better financially than they were before."

Girouard fought the good fight, but after more than a year, he gave up trying to rejuvenate his rice ponds. For the first time in almost forty years, he went to work for someone else, taking a job with the state as an agricultural expert at the AgCenter's rice research station near Crowley.

He found satisfaction in the opportunity to assist those remaining farmers, even as they faced another setback from inundation by Hurricane Ike in 2008. But he was a farmer at heart, and he missed the independence of being his own boss, making all the decisions, and putting his skills, knowledge, and instincts to use to produce his own crops on his own farm.

"It was a wonderful experience, and I wouldn't trade one year of it," he said. "I was lucky. It was an occupation I wanted to be in, and I did it. I feel like I was successful at it, and I just wish it wouldn't have had a hurricane to force me to retire."

A WAY OF
LIFE AWASH

No matter how quaint, quiet, and out-of-the-way a Cajun town seems to be, chances are it's situated on the road to someplace else. Since the widespread construction of paved highways and roads beginning in the 1930s, most towns and communities across the prairie, along the Mississippi River, and up and down the bayous of South Louisiana have been influenced by the people and commerce moving through them, while residents of once-insular locales could themselves venture out as never before for shopping, schooling, and socialization. For most of modern-day Cajun country, it's a small world after all.

That is not the case, however, in lower Terrebonne Parish. No one passes through here en route elsewhere, because there's nowhere else to go from here. This is where the highways end. And for the people of these roads, people who live quiet, modest lives amid quiet, modest surroundings, something else is ending too: their very way of life, washing away with the wetlands that once embraced them but now are disappearing before their worried eyes.

Topographically, Terrebonne Parish is an open hand. Most of its one hundred thousand-plus residents live in and around the city of Houma, at the palm; the rest are diffused down the fingers, bayous framed by strips of high ground pointing southward through the swamps and marshes to the Gulf of Mexico through communities such as Theriot, Dulac, Chauvin, Montegut, and Pointe-Aux-Chenes. Those bayou communities are largely isolated from Houma and from each other, with some eight hundred thousand acres of wetlands and open water separating and surrounding them. It's little wonder that a 2007 study by the U.S. Geological Survey rated

Terrebonne Parish among the most remote places in the country.

Along those lonely routes live Cajuns, African-Americans, and people who ascribe to assorted other ethnicities or none at all. Toward the end of most of those roads, though—farthest removed from the trappings of modern society—are Native Americans, members of the United Houma Nation.

The French explorer Rene-Robert Cavalier, sieur de La Salle, reported a village of the Houma in 1682 on the east bank of the Mississippi River near what is now West Feliciana Parish. Over the next two centuries, the Houma moved steadily southward—along the Mississippi, then Bayou Lafourche, then Bayou Terrebonne—in response to disputes with other tribes and colonial tensions between the French and the British in the region. In time, they would intermarry with Acadians, other colonial French settlers, and free people of color, and while they adopted the Cajun French language over their native Muskogean, they retained an identity as a Native American people. Today, the Houma constitute a rare stronghold for the twenty-first century survival of the Cajun French language in the out-of-the-way places of lower Terrebonne and nearby Lafourche parishes, where most of them continue to live.

Growing up in Dulac, Jamie Billiot knew the story of the Houma tribe. It is her people's story. Her family's story. Her story.

Her father, Roy Billiot, got a sixth-grade education at the Indian School that served the end-of-the-road people before court-ordered desegregation of schools in the 1960s. He made a living as a commercial fisherman; his wife, Doris, worked periodically at a local shrimp processing plant. They reared their family in a tidy, ordinary house on Shrimpers Row along Bayou Grand Caillou, close to a broad array of relatives who likewise lived off the water in the way that the Houma had done since they resettled from the farming lands into the marshes all those years ago.

As a child, Jamie Billiot learned tribal traditions from her grandparents. Her grandmother taught her about herbal remedies for common illnesses and afflictions. From her grandfather she learned how to make a blow gun and hunt small game just as members of the tribe did for centuries. They are both deceased now, and she remembers them fondly when she dances or cooks fry bread and other Native American staples at powwows.

It saddens her that her grandfather's boyhood home no longer

Jamie Billiot at the Dulac Community Center.

On low-lying Shrimpers Row in Dulac, a permanent sign warns, "Water on road ahead."

exists. Most young adults elsewhere might not get too worked up
about such a loss, but its significance for Billiot is substantial, because it isn't just the dwelling where Antoine Billiot grew up with
his nine brothers and sisters that is gone. It's Bayou La Butte, the
entire coastal island where his family lived, that has disappeared,
swallowed whole by the encroaching Gulf of Mexico.

Once a virtually solid green belt along the state's Gulf coastline, the rendering of Louisiana's coastal wetlands get speckled by
increasing splotches of blue—signifying open water—with every
update of the state map. The coastal marshes—vital as storm buf-
fers between the Gulf and more populated areas north of the coast,
as breeding grounds and habitat for thousands of native species
from shrimp to brown pelicans to alligators, and for a variety of
other ecological benefits—are vanishing at an alarming rate. The
America's Wetland coalition of scientists and conservationists es-
timates that Louisiana is losing twenty-five to thirty-five square
miles of wetlands per year. The loss between 1990 and 2000 was
documented at twenty-four square miles per year, or the equivalent
of a football field every thirty-eight minutes. And more wetlands
are disappearing in coastal Terrebonne Parish than anywhere else
in the state.

The land is washing away and inviting ever-deeper incursions
of damaging saltwater from the Gulf, due to a range of both natural
and man-made impacts, including a rising sea level, wave erosion,
soil subsidence, storm damage, construction of canals through the
wetlands for oil and gas operations, and the installation of levees
that blocked the natural overflow of sediment-rich waterways like
the Mississippi River that historically had replenished the marsh
with fresh nutrients.

As the land becomes threatened, so too does the culture whose
existence depends upon it. Billiot's grandfather and his brothers
built boats; his sisters made baskets and other traditional tribal
crafts. Those skills can still be passed on from generation to generation as long as members of the tribe remain connected, but when
she ponders the future for her extended family and the place she
knows as home, it is the loss of the coastal wetlands—exacerbated
by devastating blows from Hurricanes Rita and Ike—and other ex-
ternal pressures that trouble her.

Descendants of what had been a tribe of farmers, the Billiots, the Verdins, the Dardars, the Naquins, and other Houmas who ended up in the southernmost reaches of Terrebonne's bayou country adapted by fishing the coastal waters, hunting and trapping in the marshes, and farming what land was available along the bayous. Tribal historians point to the development of the region's oil and gas industry in the twentieth century as a tragic turn, for many of the unschooled Houmas lost their land to oil companies in transactions too complex for them to understand. The Louisiana Governor's Office of Indian Affairs has sanctioned the United Houma Nation as a valid tribe, and a recent estimate of eleven thousand members makes it the largest tribe in the state. Tribal leaders have been working since 1987 to secure recognition from the U.S. Bureau of Indian Affairs, which could make available federal services that long have been provided to other groups.

Meanwhile, times are hard, and with each storm, they get harder. With the area beset by poverty and dwindling job opportunities, and with deteriorating conditions in the surrounding wetlands making the impacts from periodic hurricanes and even lesser weather events worse than they used to be, young people are under greater pressure than ever before to leave the area and move to "the city"—Houma—or other destinations offering meaningful work and less volatile living conditions.

As that migration gains traction, and the concentration of Dulac's native population is diluted, it threatens to alter, perhaps irrevocably, the relationship those people of the bayous have had with the land and water that have been their home for generations. Jamie Billiot understands that dilemma, because she wrestled with it herself.

She had spent virtually her entire life in Dulac and the extended community along Bayou Grand Caillou until arriving at Ellender High School in Houma as a ninth-grader. For a shy girl who had led a sheltered existence on Shrimpers Row, that new window on the world was shocking.

"When I went to high school, I was exposed to other cultures and other communities and I discovered for the first time that everybody else wasn't just like us," Billiot said.

After high school, she took an even more dramatic step, enrolling at Louisiana State University in Baton Rouge.

"LSU was a big culture shock for me," she said. "It was very hard to leave Dulac, but when I was in high school, I felt like I wanted to

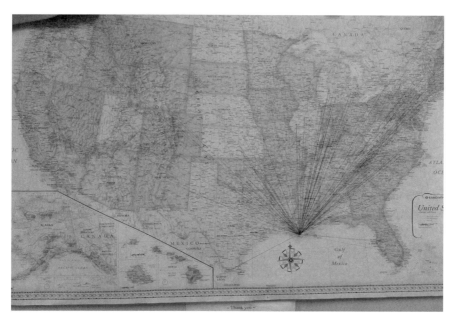

In the dormitory at the Dulac Community Center, a wall map indicates places across the United States, Canada, and Mexico from which church and school groups have come to volunteer with the storm recovery in Dulac.

leave. You get to high school and get exposed to new things, and you realize the poverty and the struggles where you are. I wanted to do better for myself. I had every intention of going off to college and never coming back to Dulac."

Her immersion into the sometimes-frenzied, culturally diverse campus life at the large public university was initially overwhelming and intimidating, but the quiet-by-nature young woman adapted in a most unexpected way, falling back on her roots.

"I found my niche in trying to educate people about my community," she said. "I was just amazed at how many people didn't realize that we existed down here. Our way of life? They had no idea. That's how I got my voice."

While at LSU, she helped to establish the first Native American students' association. After graduation, she moved to Arizona and spent three years living on a reservation near Tucson while she pursued a graduate degree in indigenous education at the University of Arizona.

After that, she returned to Dulac, determined to work with people from her own community. She arrived back home in June 2007, almost two years after Rita had wiped out the place.

"I had come back [to visit] a few times during Katrina and Rita, so I saw the devastation as it progressed," she recalled. "We were not in that immediate relief phase by then; we were in that second stage, where people were just getting back into their homes. There were still lots of needs that weren't being met."

For starters, Billiot joined the board of the Dulac Community Center, a focal point of community life in Dulac. The area is home to more than seven hundred households—about 2,500 people—and almost everyone has some reason to visit the center: computer access, children's basketball games, adult education classes, the used clothing shop, the weight-loss group, the Native American dance group, or various community meetings.

"For me growing up, there were always kids around, there were always classes happening, lots of support, socially, educationally," she said. "The community center has always been a part of my family."

Eight months later, she became the center's director.

Brimming with enthusiasm, within her first few months on the job she launched a community library to replace the public library branch that had been destroyed by Rita. Its opening coincided with the start of the new hurricane season.

Lower Terrebonne took a direct hit from Hurricane Gustav on Labor Day. Although it was a minor storm, only a Category 1, the area had lost so much of its coastal wetlands in recent years that it is left with considerably less protection from storm surges and tidal action than it once enjoyed. The result was nominal flooding for homes and other structures still at ground level, including the Dulac Community Center.

Billiot evacuated with her family to Raceland in neighboring Lafourche Parish. They returned after a few days to find that Gustav's floodwaters had deposited two to three inches of mud throughout the community center.

"We got back really quickly and started picking up things, opened our doors, and were ready to help wherever we could help," she said.

Just as things were starting to dry out, Hurricane Ike entered the southeastern Gulf of Mexico at the beginning of the next week. It took aim at the Texas coast, but it was so big and powerful that as it swept past Louisiana on September 12, far out in the Gulf, it still managed to inundate Terrebonne Parish with the worst flooding residents had ever seen—surpassing even Rita's flooding from 2005.

Billiot was at work at the community center that morning—in the rubber boots she had been wearing every day since her return from her Gustav evacuation—as the water started pushing up the Houma Navigational Canal and other smaller waterways, spilling over the banks and seeping into the center and other buildings again. Once she realized it was going to get worse before it got better, she knew she would have to leave yet again.

"The reason Ike was so bad was it was unexpected," Billiot said. "We thought we were clear. We thought we were good."

This time, the flooding reached four to five feet at the center and topped out high enough to ruin just about everything at her parents' house and many of the other homes back on Shrimpers Row. Just like that, a hurricane that went ashore 250 miles away at Galveston, Texas, undid much of the work that had gone into rebuilding from Rita, another far-offshore hurricane, fully three years earlier.

Dulac wasn't alone. The extent to which conditions had worsened throughout lower Terrebonne Parish within the span of a lifetime was staggering to those who stopped to think about it. People in Chauvin remembered how, when they were kids, their cousins from Cocodrie, fifteen miles farther down Bayou Little Caillou, would come up to stay with them when hurricanes threatened; flooding had gotten so much worse, with so much more open water where the marshes used to be, that the Chauvin residents started saying, "We're Cocodrie now." And at the small, isolated Native American community on Isle de Jean Charles, off the Pointe-aux-Chenes highway, the land loss was so extreme that local leaders began talking seriously, for the first time, about the prospect of abandoning the property altogether, surrendering it to the Gulf and moving the few remaining residents to higher ground.

As the long haul of community recovery started anew, the younger folks down the bayous found themselves facing another reason to consider moving away. The young director of the Dulac Community Center recognized what was happening. While she could make sense of it, it still frustrated her.

For members of the Duplantis family of Grand Caillou, just north of Dulac in lower Terrebonne Parish, flooding from Hurricane Rita dumped not only water but mud and marsh grass into their home.

Tommy Higginbotham, right, helps Clark Duplantis remove living room furniture from the Duplantis family's flooded home.

"The elderly and the people who have been there all their lives, they still want to fight," Billiot said. "Within the younger generation, it's harder to find people who want to invest in the community.

"It is scary. If we don't do something, if somebody doesn't speak for our people, if no one fights for our people, then my daughter won't have anything that's relatively similar to where and how I grew up. That's a scary thought. We should protect our land. This is hard for us."

Billiot was keenly aware of how the problems facing Dulac were beyond the local residents' power to solve. With her informed perspective after seven years in the outside world, though, she despaired of how difficult it was for people elsewhere to understand the personal impact of the wetlands crisis on the people living down the bayous, whether from a storm surge that flooded a shrimper's uninsured home or the encroaching saltwater that consumed the land where cattle once grazed or people once lived or loved ones were buried—or, most recently, a catastrophic offshore oil spill which menaced Gulf waters, adjacent inland waterways, and those coastal wetlands that hadn't yet been washed away, thereby endangering as well the only livelihood most of the area's people knew.

"It's hard to make people on the outside understand about our coastal land-loss issues—how it threatens our culture, how it threatens our language, how it threatens our way of living," she said.

But some have gotten the message and have been moved by it. Ever since Hurricane Rita, out-of-town volunteers have landed by the busload in Dulac (as elsewhere across South Louisiana) to help with clean-up and rebuilding. There have been church members, high school and college students on spring break, civic organizations, and others from throughout the United States, often for weeks at a time. And no group's experience at the end of the road has been more profound than that of Janet Grigsby's students.

A sociology professor at Union College in Schenectady, New York, Grigsby escorted a group of her students to New Orleans in 2006 to volunteer on two Hurricane Katrina recovery projects. Upon their return, she developed an academic course around the volunteer work in Louisiana, which she rightly sensed would be ongoing for some years. She soon came to realize, though, that there would be more value in getting students into the communities affected by disaster and prodding them to get to know those people and the way they lived, than in merely working on a construction crew to build a house for someone they'd never even meet.

"After the hurricanes in 2005, the most that people got out of it was 'the levees, the levees, the levees' and 'why didn't FEMA do what it was supposed to do' and 'why didn't the Corps of Engineers build the levees right' and 'why are people living there in the first place' and 'why the heck would people go back to live there,'" Grigsby said. "That's like: how superficial, such a biased and unproductive way to look at it."

With encouragement from Kerry St. Pé of the Barataria-Terrebonne National Estuary Program, Grigsby added a second week of duty in Dulac to the curriculum. With that, the course transformed from field trip novelty to meaningful, enlightening, and deeply enriching experience.

"To me, that's almost the most important part of the course," she said. "New Orleans is fine and they all think it's cool to go to the Big Easy and all that stuff, but to really understand South Louisiana, you need to get out to the wetlands, because the loss of the wetlands is so critical for New Orleans and the wetlands are so crucial for the country as a whole."

Grigsby accepts fifteen to twenty students for the course every year. They meet periodically during the fall, establishing an academic context for what will follow, then travel down to Louisiana together for two weeks during the college's winter break. Upon their return to campus in January, the students write ten to fifteen page term papers on a topic relevant to the course and what they experienced doing hurricane recovery-related community service. The activity concludes in late January when they present a symposium for the student body to relate what they've learned.

"The students have almost as much culture shock, if not more, as if they were going across the border somewhere," Grigsby said. "They're startled by New Orleans because it's not like anything they've seen up north, and then they're even more startled down in Dulac where, first off, there is the deeply entrenched poverty. Most of my kids are from pretty well-heeled backgrounds. They have not seen communities where so many people live in what they would call just trailers, and they haven't seen people who live in the physical conditions that many of the houses that we've worked on have been in.

Union College students Carl Winkler, Sarah Yergeau, Molly Head, and Katie Ferrara prepare to start gutting a flooded house in Dulac.

Union College students Carolyn Stonerook, Sarah Yergeau, Chelsea Charette, and Jasmine Maldonado [on ladder] repaint a flooded home in Dulac.

"We spend a lot of time in reflection sessions at night. They write journals and we talk constantly about, why is it like that? What does that mean? What are your own feelings about this and why are you feeling that way? How do you feel about the people? We try to get them to walk in other people's shoes and get them to understand it."

When the students get back to Schenectady and begin to process what they saw, they invariably marvel at how much they learned—and how they didn't know what they didn't know.

"It always means a lot to volunteers, but I think it especially means a lot to my college student volunteers if they get a chance to meet and get to know the owners of the homes they're working on," Grigsby said. "They like to be able to put a face with the homes, so it's not just a house they're building but it's somebody's home."

By and large, residents have welcomed the assistance and embraced the visitors. Whether they were floating new drywall in someone's storm-damaged house or picking their way through their first-ever boiled crabs after a hard day's work, students found it didn't take much to get the locals talking about the joys and hardships of life on the bayou.

"I love when college groups come down," Billiot said. "You can really tap into some activism. And Dulac is a very welcoming community. Our people love to share with them. They want to share their stories."

The students also spend a day on a wetlands restoration project, such as a site at Port Fourchon where they plant grass plugs in a patch of wetlands that is being reclaimed by the estuary program. It is dirty, wet, monotonous work, usually done to the accompaniment of gnats and mosquitoes, and their teacher never hears them complain so much as they do that day. She is heartened, though, to catch them occasionally scrutinizing the sandbar they were restocking with vegetation: "They were calculating: 'How many football fields is that? Have we made up for the one football field every thirty-eight minutes, or are we still behind?' That's the kind of sensitivity thing that I'm just so pleased about."

Several of the students have been so moved by the experience that they have returned to Louisiana on their own to volunteer, or work full-time, on continuing recovery efforts.

"One of my goals for the course is that they'll go down there

Union College students Christian Ramos, Meredith Adamo, Kaleigh Ahern, and Rachel Feingold plant marsh grass at Port Fourchon.

and they'll do a lot of good work but they will also learn a whole lot, and they will learn things that not too many people back home know but that are important things that people should know," she said. "It's all part of their responsibilities to find ways to tell the story of the people they have worked with."

And every spring, when she solicits applications for the class, Grigsby always says, "God willing and the storms don't come again, we're going back."

DEAD AND GONE

Sue Fontenot had all the charm of a snapping turtle.

She smoked big cigars, drove a pickup truck, cussed like a roustabout, was highly opinionated, and had a fiery impatience with bureaucratic ineptitude. At a time when Acadiana's bar and bench were still very much a boys' club, Fontenot elbowed her way into the fraternity after earning her law degree in 1971. She went on to make a name for herself as one of the few women elected in the 1970s to a state district judgeship, with her tenure on the Fifteenth Judicial District Court bench bracketed by a successful law practice. As a judge, she was known to hold court until late in the night, because she could—and woe to the attorney who ever entered her courtroom unprepared. She was a force of nature, one of those characters long remembered—for better or worse—by all who encountered her. But she was loyal to her friends and clients, devoted to her family, and affectionate toward the place she called home.

Home, to Sue Fontenot, was Mouton Cove. A settlement of a few farm houses amid pastures and woods, a few cows, and a few crops—it's hardly a speck on the Vermilion Parish map, and lucky to rate that much. Mostly, it's one of those places people don't even notice as they drive past, on the way to somewhere else.

Fontenot had known Mouton Cove most of her life. Divorced, her children grown, she lived alone in what had been her parents' home. It was just a few minutes' drive from her law office in the parish seat of Abbeville, but the serenity of its wide open spaces made it seem a world away.

Like many of the communities bordering the coastal marshes of lower Vermilion Parish, Mouton Cove stood in harm's way when hurricanes would approach, and Fontenot would evacuate

74

at the first sign of trouble. Flooding wasn't a concern, because her house was well-situated on a patch of high ground. Rather, it was the inevitability of the power outage that followed any hurricane landfall—from several days to a week or more—and the resultant, if temporary, loss of creature comforts that prodded her retreat to points north. For Hurricane Rita, she drove seventy-five miles to Ville Platte, where she spent five leisurely days with friends, playing cards, watching TV, eating too much, and otherwise passing the time until it was safe to return home.

A few days in, though, a phone call from a friend back in Vermilion Parish brought ominous news: Mouton Cove was under water.

Fontenot didn't fret about her house. If it was flooded, she knew, she could always relocate to town and live out of her law office for awhile. Her thoughts instead turned to Esther.

A crossroads spot a few miles south of Mouton Cove, Esther is home to the Catholic mission chapel of St. James, the church she and her brothers had known since their childhood. Her parents and her youngest brother were buried in the small country cemetery behind the small country church, and her recently deceased brother James had been laid to rest in the cemetery's mausoleum.

She knew enough to worry, because when hurricane season comes to South Louisiana, there is no guarantee that the dearly departed will rest in peace.

"What about the cemetery there?" she asked from her exile.

"I'm sorry to tell you, but it's awful," the caller informed her. "The graves have come out."

As soon as local authorities opened the roads, Fontenot made her way back to Vermilion Parish. She drove past her house and went straight to the cemetery at Esther. She was relieved to find the graves of her parents and her younger brother Brad undisturbed in the soggy ground, but at the cemetery's mausoleum, her darkest fears were confirmed: James was missing.

She was not alone in her dismay. It was a bitter irony for thousands of South Louisiana residents that although only one life was lost to Hurricane Rita, hundreds already dead and buried were lost—literally—in the flooding brought by the storm.

Damage inflicted by the worst storms to hit coastal Louisiana is seldom limited to destroyed homes and washed out roads. With wave action, swift currents, and inundation, storm surges and pro-

longed flooding brought on by major hurricanes can dislodge—by
the hundreds—caskets from their burial sites in cemeteries where
the water table isn't very far below the soft ground. Thus, the hardships endured by many hurricane victims are compounded by the
grim discovery that the remains of loved ones have been disinterred and washed away by floodwaters.

In places like the compact town of Delcambre, where the water rose, did its damage, and then receded, evacuees returned to
find macabre scenes of caskets littering yards and sidewalks and
perched atop cars and trucks. About twenty miles away, the road
leading to the isolated St. Mary Parish settlement of Louisa was
blocked by twelve or so caskets, some of the sixty that had floated
away from St. Helen's Cemetery there. In such places where caskets
were scattered in plain sight, sheriff's deputies, public works employees, and other local authorities soon set about retrieving them.

The impact was all the worse in the communities strung out
across the lower reaches of Cameron Parish and, to the east, in Esther, where caskets dispersed into swamps, marshes, or open water. The unlikely spectacle of a metal casket parked on a sidewalk
or lying in a roadside ditch a block or two down the street from
a graveyard was unsettling enough to residents in some locales.
Much worse, though, was the punch-in-the-gut shock awaiting
family members who arrived at decimated cemeteries to find a
spouse or parent's empty grave site, their pain soon intensified by
the despair that accompanied the realization that the loved one's
casket wasn't just uprooted, it was *gone*.

Some of them were never found. But most of them were recovered—weeks, months, even years later, thanks to a core group of
unflappable volunteers who never made the hurricane rescue highlight reels but were no less heroic, fending off alligators, snakes,
insects, and fatigue far into the marsh and out of the limelight, day
after day for months on end.

The exhaustive Cameron Parish effort was headed up by Zeb
Johnson, founder and owner of Johnson Funeral Home in Lake
Charles and a deputy coroner for neighboring Calcasieu Parish.

Three days after Rita's landfall, the Calcasieu coroner's office
was asked by the FBI to assist in identifying some bodies that had
turned up in Cameron Parish. Reliable information about the extent of Rita's impact across coastal Cameron Parish was still hard

to come by, since much of the parish remained flooded and cut off from the outside world, but to that point there had been no reports of storm-related deaths, so the FBI's inquiry met with some initial uncertainty. Once the local authorities determined that the bodies in question were not Rita-related deaths but corpses that had been extracted from cemeteries by hurricane floodwaters, "we thought that we were talking about some isolated cemeteries that have some flooding problems during the heavy rains, things like that, just localized flooding," Johnson said.

Initial forays into the flood zone told a different, much more harrowing story: all but two of Cameron Parish's forty cemeteries had been breached.

"Tombs were out of the graves. Most of the tombs were broken open. Many of the caskets had begun a journey through the marsh to places unknown," Johnson said. "We discovered that this was going to be a much bigger problem than anybody had anticipated."

Because the ground is sandy or mucky and barely above sea level, caskets buried in the cemeteries of Louisiana's Cajun coast typically are installed in surface vaults. The concrete shells are set only a few feet into the ground, with heavy lids protruding just eight inches above the surface. Many vaults were uprooted by Rita's floodwaters and pushed far inland, with caskets still inside. In hundreds of other cases, vault lids were dislodged and caskets popped out of the vaults and floated away.

By December, FEMA was estimating that about 1,300 caskets had been dislodged from burial sites throughout the state, eight hundred by Rita in Acadiana and five hundred by Katrina in the New Orleans area.

This had happened before, but never on such a massive scale.

"We heard stories during Audrey of this happening in Cameron Parish," Johnson said. "But, first of all, there had not been as many burials then as there are now. You're talking about a fifty-year span of time. There is a progressing number of people that have been buried since then." For Rita almost fifty years later, "the cemeteries were more exposed and they had more people in there, so naturally there were more bodies missing."

Record-keeping for such a disaster-within-a-disaster was primitive back then. So were recovery methods.

"During Audrey, they didn't have the EPA and the DEQ and

Kristi Mock sweeps dirt off an unmarked grave at Our Lady Star of the Sea Catholic Church in Cameron, June 20, 2007.

every other federal agency flying around in helicopters spotting things and doing things," Johnson said. "Most of that was done by boat, by horseback, or things like that. A number of people told us that for Audrey, they went out on horses and pulled some of the vaults and caskets back and reburied them themselves. We had no accurate recording of how many people were missing other than family members saying that they were told that their grandfather had left during Audrey, things like that."

FEMA was widely ridiculed for its insufficient response to the dire back-to-back hurricane disasters that befell Louisiana in 2005. For all the shortcomings that FEMA demonstrated in tending to survivors of Hurricane Katrina and Hurricane Rita as the crises unfolded, it was well-prepared for a more morbid but still necessary duty: dealing with the dead. In each event, it activated a Disaster Mortuary Operational Response Team to help local authorities recover, identify, and bury storm victims and the disinterred remains from flooded cemeteries. Following a procedure that the government established in the 1990s in response to large-scale transportation accidents such as airplane crashes, FEMA mobilized teams of funeral directors, coroners, dental professionals, X-ray technicians, and support personnel to help recover and identify bodies, prepare them for burial—or reburial—and locate and notify relatives of the victims.

Johnson and some of his colleagues from the Lake Charles area spent two and a half weeks in New Orleans immediately after Katrina as members of FEMA's team there, recovering and tending to some of the 1,464 bodies left through the city and its suburbs in the storm's wake. They returned just in time for Rita.

For D-MORT volunteers who worked both hurricanes, Katrina "was a totally different issue, because those bodies were in houses and on the side of the road, and here these bodies were out in the marsh," Johnson said. "It was a big problem."

Most of Cameron Parish is either open water or marsh, and little of the land that does exist rises above sea level. Consequently, the caskets and burial vaults that floated or were pushed out of cemeteries across the parish ended up in lakes, ponds, other waterways or remote, inaccessible marsh areas. It took a few weeks for Johnson and other authorities to sort things out down there. They remapped every cemetery in Cameron Parish—all forty of them—making a census of gravesites they found intact and plots that were empty.

The missing bodies totaled about 340.

"You've got several different problems," Johnson said of the challenge that awaited his team. "You've got not only recovery, you've got identification. Then you've got recasketing, because most of the caskets were damaged, and then you've got the reburial. So you've got four things that you had to do."

Just finding the missing was a massive ordeal. Some caskets that floated away in the surge settled high in the branches of trees. Others landed randomly in yards, open fields, or roadside ditches as the floodwaters receded. Many caskets and vaults sank into the water and mud in the vast expanse of marshes and swamps, where the recovery team relied on spotters in helicopters, airboats, and marsh buggies to locate them. In some cases, older wooden caskets just fell apart after prolonged exposure to the elements, and remains had to be gathered by hand.

Once the recovery from Rita began to gain traction in and around Cameron Parish, the search was aided by the volume of helicopter traffic overhead as workers were ferried to job sites in Cameron or offshore.

"During Rita, we had all these helicopter companies flying back and forth," Johnson said. "People knew we were looking, so these pilots would spot things and they would GPS them for us and we would go and find them."

Some caskets could be brought in from the inland waterways by boat. Marsh buggies with their high profile and tank-like tracks provided more effective, if slower, access to caskets in the wildlife refuges and other marsh and swamp areas that account for most of Cameron Parish's surface area. And others were hauled out of some of the most remote locations by helicopter, after workers on the ground wrapped them in chains and connected them to a line dropped from the hovering aircraft.

For the locals involved in the effort, it was merely exhausting. The occasional casket-recovery assignment deep into the marshes bordered on other-worldly for dutiful but wide-eyed National Guard troops deployed from throughout the United States to storm-ravaged coastal Louisiana. Wherever in the marsh the caskets turned up, the recovery teams usually could reach them only by negotiating hip-deep mud, voracious mosquitoes and gnats, and the occasional water moccasin and alligator.

The recovery team set out intending to rebury all of those washed away by Rita, but three out of every four bodies they recovered contained no identification. Details about the deceased are supposed to be written out and placed in a glass capsule known as a memorial tube, affixed to one side of a casket. Many funeral directors fail to do so, however, and in the case of flood-disturbed caskets, those papers that were filled out in ink instead of pencil were ruined anyway. For Johnson and other funeral directors committed to compassionate care for the deceased and their survivors, dumping unidentified bodies of local residents into some mass grave was not a preferred option, so they would examine the remains in search of unique characteristics—how they looked, what they were wearing, and what personal objects the caskets contained. A family assistance program was set up for people to provide information about missing caskets and the relatives buried in them.

As they were recovered, many of the bodies were sent to Carville, where FEMA had set up a temporary morgue after Hurricane Katrina, and forensic experts there X-rayed and studied the remains in further attempts to identify them.

acuees from Texas and southwestern Louisiana clog the eastbound lanes of e Interstate 10 bridge over the Calcasieu River at Lake Charles on Thursday, t. 22, 2005.

fton Hebert, Cameron Parish's first full-time emergency preparedness di- tor.

Visitors to Grand Pré, Nova Scotia, are drawn to the memorial cross mark the site of the Acadian deportation.

A visitor in Grand Pré, Nova Scotia, examines a memorial plaque listing nam of Acadian families deported by the British, beginning in 1755.

Rita slams into Lake Charles, September 23, 2005.

Aerial view of flooding in Cameron, September 25, 2005.

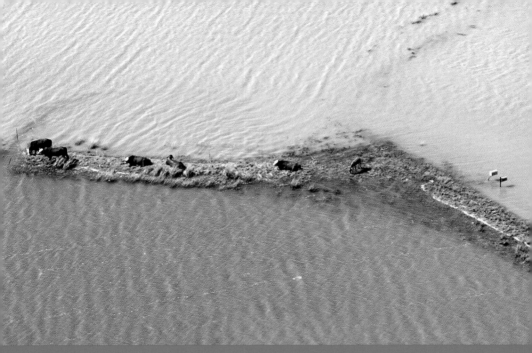

Cattle stranded by Hurricane Rita floodwaters in Vermilion Parish.

Flooded cemetery and flattened trees in Creole, the day after Hurricane Rita's landfall.

...nnants of homes and camps, including boats such as this, were pushed miles ...nd by Hurricane Rita's storm surge and deposited into the sugarcane fields ...beria and Vermilion parishes.

...er Delcambre was swamped by Hurricane Rita's storm surge, only the ...nt steps of this mobile home were left.

For the Duplantis family of Grand Caillou, just north of Dulac in lower Terrebonne Parish, flooding from Hurricane Rita dumped not only water but m and marsh grass into their home.

The town of Erath, including the Catholic church and cemetery and near sugarcane fields, inundated from Hurricane Rita's storm surge.

yor George Dupuis looks over an empty field north of Erath where, in the
ermath of Hurricane Rita, planners recommended relocating the town
ay from its flood-prone location. June 21, 2006.

rricane Audrey memorial in front of Our Lady Star of the Sea Catholic
urch, Cameron.

Charlie Theriot, 95, and his wife Macilda, 91, relax in temporary quarters Grand Lake as they wait for a new mobile home to be readied on their hon site in Grand Chenier, June 20, 2007. Macilda died two weeks later.

Nothing but flowers where a neighborhood in Cameron Parish stood befo Hurricane Rita.

"What they did down there was, they would help us to give us some clues about clothing description, male or female, personal items that were in the casket," Johnson said. "By doing that and contacting the local funeral homes, and through a process of elimination and intrinsic things, things that were in the casket that we were able to identify, we ended up identifying about 85 percent of the bodies and returning them to their cemeteries."

Two of the caskets recovered in Cameron Parish were particularly oddly weathered and didn't match any of the records the team had compiled about the missing from parish cemeteries. Eventually, the investigators figured it out: those had been cast out by Hurricane Audrey way back in 1957.

"Most of the families were not even available," Johnson said. "It was just through pure luck that we were able to identify those two bodies and we actually returned them to where they were buried."

All of the local funeral homes pitched in with the reburials in the months after Rita. The real challenge was in the recovery of the caskets from Cameron Parish's wide-open spaces, with only a few volunteers who were committed to the effort but hampered by a lack of equipment and lack of resources.

"For guys to go out and take a beating seven days a week during that recovery process, we just couldn't get many people who wanted to go out and do that," Johnson said.

For the handful that did, the ordeal lasted nine grueling months and exacted a tremendous physical and emotional toll. Looking back on the impact the recovery experience had on them, the funeral service professionals and others who took part had a saying: many went out but only a few returned.

"None of us were really the same after we came back from that," Johnson said. "Our lives had changed—it was like we lost that whole year out of our life. We abandoned our businesses, we abandoned our families. We were gone. We've got guys out in the marsh, trying to make recoveries, and they've got their homes that have been damaged, so they're coming in at night trying to take care of that. We had a pretty high sense of dedication with these guys who went down there to do what they felt was the right thing to do. They really sacrificed their personal lives and their business lives to be down there doing that."

Passers-by look at coffins that were placed on a flatbed truck after they flooded out of their graves at Our Lady of the Lake Catholic Church cemetery in Delcambre.

Graves in a Delcambre cemetery were disturbed by Hurricane Rita floodwaters.

And when Hurricane Ike hit in 2008, they did it all over again.

"We at first thought Ike was a very limited storm with a limited amount of damage," Johnson said. "Within the first twenty-four hours, we had found several caskets in Lake Charles that came from cemeteries deep into Cameron Parish. We knew that we were going to have some problems."

As Ike's floodwaters washed over the cemeteries of Cameron Parish, they dislodged about two hundred caskets. That was only slightly more than half of what Rita had displaced, but it didn't mean the recovery effort was going to be any easier this time. Most had been disinterred by Rita, recovered and reburied, only to float away once more when Ike hit. This time, the caskets were more dispersed. Many were new, having been replaced after being lost and recovered in Rita.

"They were much better caskets, much more durable, so they became much more buoyant and they floated great distances," Johnson said. "They became like boats. We found them more widely dispersed. We found four caskets that literally traveled side-by-side and they ended up twenty-three miles away from the cemetery."

Seven caskets unearthed from Cameron Parish cemeteries by Ike were pushed by wind and tides across the state line into southeastern Texas. Recovering those bodies posed more of a challenge than usual.

"It required the use of airboats, marsh buggies, and different things to get them out of there, and those resources are not normally readily available in Texas," Johnson said. "We had to send our equipment over there and our manpower over there, and just the logistics of getting them over there and getting them back became a real big problem."

Meanwhile, between Rita and Ike, the federal government had changed the D-MORT operating procedures.

"They no longer assisted us in recovery. They would only assist us in identification," Johnson said. "We pretty much burned everybody up after Rita, so we had a very, very hard time in finding people that were willing to participate and help us with the recovery. Instead of having about forty people involved in that, we ended up with about five people that actually assisted on a daily basis in the recovery of the caskets from Ike. That made it much more difficult for us. It was much more time-consuming. It was much more frustrating. Guys had to work eighteen and twenty hour days, seven days a week. It just became a very, very difficult issue."

"Going into this, all of us knew there was no reimbursement coming. We knew we probably were not going to get paid for this. Fortunately, there were some donations that actually paid for the replacement of the vaults and the work in the cemeteries. Most all of the people that worked down there volunteered their time and their services. They actually were never paid for it. We feel like this is what we should be doing. This is what we work for, what we train for."

By the one-year anniversary of Hurricane Ike in September 2009, eight to ten caskets uprooted by that storm were still missing. About twenty of the 340 lost to Rita remained unaccounted for, four years later. The formal search had long since been curtailed, but the recovery team wasn't giving up. It couldn't. This was personal.

"The families were very concerned about the return of the bodies," Johnson said. "We became very emotionally attached to those people. We talk to them frequently."

The only two caskets lost from the Immaculate Conception Cemetery in Grand Chenier were a husband and wife that apparently floated off together. After three and a half years, they were still missing. When business was slow back at his funeral home, Johnson would head down to Cameron and spend hours driving around, standing atop his truck, and looking through the marsh, hoping to spot something he might have missed the last time. Relatives were appreciative of the efforts, even though the couple hadn't been found. Johnson would see them once or twice a month, and they would ask him "Any luck? Any luck?"

National Guard personnel flying over the marsh in a helicopter eventually discovered the wife's vault, and a local contractor with an airboat made the recovery and brought it in. On the day that family members were moving back into their home after making repairs from Hurricane Ike eight months earlier, Johnson paid them a visit.

"When I saw them, they said, 'This is a great day, because we're moving back to our house.' I said, 'Well, I have more great news for you. We found your mother,' " Johnson recalled. "We were able to rebury her in another cemetery. At the same time, we made a place for their dad to be buried, because I told them I won't rest until we find him."

Johnson was similarly moved by the plight of a terminally ill Cameron Parish woman whose mother had washed away in Rita. More than eleven months later, on Labor Day 2006, a search crew found the casket out in the marsh.

"She was dying with cancer and she had told me she probably would die before we found her mother," Johnson said. "When I walked to her house and gave her her mother's wedding ring, I didn't even have to tell her anything. I said, 'This is your mom's wedding ring,' and she knew we had found her mom. For all the work we did, all the sweat, all the tears, those moments like that are priceless to us. There's no amount of money that can pay for the feeling that you get, being able to do something like that for a family."

Thirteen months after Hurricane Rita devastated Cameron Parish, a stunning monument of black and gray polished stone, six feet tall, was dedicated on the front lawn of the parish courthouse. With no one killed in the storm to memorialize, the marker instead commemorates the loss of the more than 340 bodies scattered by the hurricane's floodwaters. The inscription reads, in part: "It has been a long journey for them, their families, and those that recovered them. May they again rest in peace."

But some remain missing, and Zeb Johnson cannot rest until they do. Every several weeks, Johnson's business takes him on the long drive from Lake Charles back down to lower Cameron Parish. Binoculars at the ready as he drives along the wide stretches of marsh and open water, he is compelled to pull over and look around, again and again. "I want to check this out," he'll tell himself. "I want to see if there's anything here. I know we found a few here. Maybe there's something we missed." Setting out on what should be a trip of ninety minutes to two hours, he'll be gone all day, still looking for those missing caskets.

"I guess I'll never be satisfied until all of them are recovered," he said, "but I know that, realistically, that's probably not going to happen."

In neighboring Vermilion Parish, meanwhile, Hurricane Rita dislodged 252 caskets from cemeteries throughout the parish's southern reaches. It took fifteen months to track them all down and recover them; the last reburial took place just days before Christmas 2006. Ike dislodged another fifty-two caskets in 2008, and all of those were found and reburied, too.

The greatest disruption occurred in Esther, where caskets were sucked out of the ground or blasted out of mausoleum crypts and washed deep into a swamp.

It was four days after Rita's landfall that a determined Sue Fontenot made her way back to the Esther Community Cemetery. It wasn't in her nature to allow herself to be overcome with emotion, but as she stood alone in the shadow of the St. James Chapel, with her brother James' empty mausoleum vault to one side, her parents' graves to the other, and the ground still soggy from the hurricane's storm surge, memories washed over her.

Tenes Fontenot had grown up in the cotton fields of early twentieth century Evangeline Parish. He couldn't read or write, but he knew hard work, and he could play the fiddle. Finding work in Vermilion Parish as a young man, he courted Maudry Marceaux, a bright young woman who had graduated from high school at a time when few young women in the Cajun countryside did so. The couple eventually married, settled in the small town of Kaplan and had four children. When Sue, their only daughter, was eleven, the family moved down to Intracoastal City, a much smaller, more remote settlement in the middle-of-nowhere marsh of the parish's southern reaches, at the end of the road heading south from Abbeville.

Today, Intracoastal City is a noisy, bustling crush of workboats, machine shops, and affiliated businesses, a vital staging area for Louisiana's offshore oil industry. The place that Fontenot remembered from her childhood, though, couldn't have been more different. Typical of Louisiana's isolated bayou country in the years before modern roads, modern communication, and modern jobs nudged it into the twentieth century, it wasn't a city at all, despite its name. The people lived off the land, and the land in this case wasn't even terra firma. Some, like Tenes Fontenot, claimed plots of mucky acreage for rice farming. Others, like the Fontenots' nearest neighbors, the Leges, were trappers, plying the bayous and marsh grasses for nutria, muskrat, and other critters whose pelts could fetch a modest sum. As the oil field encroached on the farming community, though, the Fontenots eventually moved a few miles up the road to Mouton Cove.

Through it all, Sue had a best friend, confidante, and mentor in her brother James. As children, they sang in three country churches in the area, including their home parish of St. James in Esther, where they started the choir. James was a particularly talented musician, equally adept at clarinet, oboe, organ, and bagpipes. He was an organist and choir director at St. John Cathedral in Lafayette, and he maintained the pipe organ for the nuns at the Carmelite Monastery across town. He spoke seven languages, including Latin. He was an attorney by profession, specializing in property and succession law, and he was elected a Louisiana state senator at the age of twenty-seven, unseating Louisiana's most famous Cajun politician at that time, Dudley "Coozan Dud" LeBlanc.

Like his sister, he also had an intense interest in Louisiana's French connections. Their Fontenot ancestors had arrived in Louisiana directly from France and the pioneer Marceaux brothers had gotten off the boat in Mobile, Alabama, before making their way to the bayous. The family's Cajun pedigree was through their maternal grandmother, a Broussard whose roots extended to the pre-deportation Acadian homeland of modern-day Nova Scotia.

James was a key figure in CODOFIL, the Council for the Development of French in Louisiana, founded in 1968. The agency teamed up with Louisiana Public Broadcasting in 1981 to produce "En Francais," a series of thirty minute public-affairs programs in French, and James, a long-time member of the CODOFIL board of directors, hosted the show. Launched with documentaries provided by Canada and France and interviews with local and visiting French-speakers, the program eventually expanded to cover cultural events and produce original documentaries on Louisiana's French heritage.

Although the sophisticated James was two years older than his brash, rough-and-tumble sister, Sue always thought he would outlive her because she was so *canaille*, as the Cajuns say—such a rascal—not to mention a smoker. As liver cancer was cutting short James' life at the age of fifty-nine, he took it better than she did.

Recognizing that he was in failing health, James visited Vincent's Funeral Home in Abbeville to make his own arrangements. He selected a simple pine casket and warned the funeral home staff, "When I go, my sister's going to come in here, and she's going to tell you, 'I'm not burying my brother in that piece of crap!' She's going to want something else, much more expensive. And you tell her if that's what she wants, that's fine, as long as she pays for it."

A haunting, familiar sight following Rita and Ike: coffins pushed by floodwaters into the marsh. This one was unearthed from Holy Family Cemetery No. 2 in Dulac.

The Hurricane Rita memorial in front of the Cameron Parish Courthouse.

He died on July 22, 2004. When Sue showed up at Vincent's to finalize his funeral arrangements, she asked to see the casket her brother had picked out for himself. "It was like in the westerns—a pine box," she recalled later, laughing as she told the story. "I said, 'I'm *not* burying my brother in that cheap piece of crap!' It just came out. He predicted accurately how I would react."

Fourteen months later, after Hurricane Rita blew past Vermilion Parish, the beautiful cherry casket she had gladly purchased for her brother was nowhere in sight, his mausoleum vault empty.

Funeral director and longtime family friend Russell Frederick surveyed the area and thought he saw what might have been James' casket somewhere in the distance. It was hard to tell, though, and the site would be difficult to reach. Sue knew someone with an airboat, but Frederick warned her that the wooden casket might not survive a harsh towing through the swamp behind a speeding airboat. She knew local contractors who had heavy equipment that might have been useful, but it was all disabled after getting flooded by the storm surge. The thought of her brother's remains abandoned in the swamp became all-consuming. Sue couldn't sleep at night and couldn't do anything in the daytime but try to find some way to reach the casket and bring James back to the cemetery in Esther.

A call to the office of Gov. Kathleen Blanco, to see what resources might be available for a recovery operation, ended in frustration when the staffer started talking about services for Hurricane Katrina victims and seemed unaware of the dire situation facing those who had just endured Rita's impact in the Acadiana region. Emotional and sleep-deprived on top of her usual no-nonsense attitude, Fontenot hung up.

In her desperation, she called an acquaintance at a television station in Lafayette, and the reporter eventually came out and did a story on her plight for the evening news. The next afternoon, as she sat in a daze in her law office, someone called her to say that a group of soldiers was being dispatched the next morning to retrieve James' casket.

Frederick met her at the staging area the next morning and stayed with her as members of a Utah National Guard unit set out into the swamp.

"Those men were knee-deep in snake-infested water," Fontenot said. "There were about twenty of them, I guess. I knew they were frightened, between the cottonmouths and all that. They got the casket, which we believed was James. It took about two or three hours, and they laid it down. Russell said, 'Just stay here.' He opened the casket, which was easy to do. It was broken on the top. He opened it, and he nodded at me, 'It's him.'"

The feeling of relief was palpable. She told the soldiers, "You will never know how grateful I am and I will never forget each and every one of you. I know this is strange territory for you. It took you to come from Utah to get my brother." And she gave each of them a hug—and a cigar.

Fontenot had her brother reburied near their parents, in the ground this time, under a sturdy granite slab. It was inscribed with a Latin phrase he had favored: "Quomodo cantabimus canticum Domini in terra aliena?" Taken from Psalm 137, it can be translated as, "How can I sing praise to the Lord in this strange land?"

"He was, I guess, agnostic," she said. "He knew more about the Bible and the Catholic religion than most priests and nuns do. . . . He wanted so much to believe, but as often is the case with extremely intelligent people, they have to have answers. They have to have proof. I kept telling him, 'You're missing the point. It's called faith.' But anyway, he was precious and I know he's in heaven."

Fontenot continued to work as a lawyer in Abbeville, where she was either "Miss Sue" or "Judge" to just about everyone. She took pride in the work of her two children, Jean-Paul, a physical therapist, and Ahna, a lawyer. Like a veteran schoolteacher greeted by a long-ago student all grown up, she would revel in the occasional contact from those who had passed through her criminal docket as youthful offenders years earlier and outgrown their wayward tendencies, invariably recalling her stern admonitions from the bench that helped to scare them straight.

She was attacked in her isolated home at Mouton Cove one night in 2007 and, true to form, fought off the intruder. The subsequent all-points bulletin issued by the Vermilion Parish Sheriff's Office advised deputies to be on the lookout for a man with bite marks and eye injuries.

Fontenot died in the summer of 2008 at age sixty-two, after her own battle with cancer. It was no coincidence that she arranged to be buried between James' new grave site and the tree line beyond the cemetery. Thus could she have the final word, for all eternity: if any hurricane ever again should dare to try and wrest her brother from his final resting place, it would have to do so over her dead body.

HEARTBEAT OF
A CAJUN TOWN

Erath is a small town surrounded by sugarcane fields and Cajun character.

At a time when Acadiana's native tongue is dying out with the older generation that grew up before the age of television, Erath boasts a disproportionate share of Cajun French speakers, attesting to Vermilion Parish's description as "the most Cajun place on Earth." Even the stop signs are bilingual, offering motorists the optional directive to "arret."

The houses are small and quaint, with chickens running around some of the yards. Residents stage a five-day festival every summer to celebrate the Fourth of July, replete with pageants, concerts, fireworks, and spirited water fights between local fire departments. The town's most famous resident is the irrepressible musician D. L. Menard, whose up-tempo drinking song "The Back Door" rivals the classic waltz "Jolie Blon" for honors as the "Cajun national anthem."

For all its Mayberry-on-the-bayou charm, Erath might best pinpoint its heartbeat just down the street from the Acadian Museum, the library, and City Hall, at 202 South Kibbe Street. Nothing defines Erath like Champagne's Supermarket.

It's where the Girl Scouts set up to sell cookies every winter, where the youth football league conducts its registration every August. It's the business that donated and maintains the big Erath High School message sign in the middle of town. Employees treat customers like family, just like the way the owners treat the employees. If you live in Erath, Champagne's Supermarket is the place where everybody knows your name.

When Hurricane Rita flooded 90 percent of the homes and businesses in town, Champagne's Supermarket was wiped out, like just about everyone and every place else. The store took on two feet of water, and everything was ruined. Erath was in for a long slog, and so were the members of the Champagne's team, many of whom lost their homes as well as their jobs to the flood. In the difficult months that followed, the fate of the devastated town of Erath would prove to be inextricably tied to that of its iconic local supermarket. That relationship placed a substantial sense of responsibility on the shoulders of store manager Ricky Luquette.

It isn't much of a stretch to say that Luquette has been working at Champagne's Supermarket all his life. It was back in January 1968 that the store was opened by his uncle, Lester Champagne, and a few other family members. Luquette went to work there a month later, stocking shelves, making deliveries on his bicycle, and keeping the place tidy. He was twelve years old.

Champagne's was founded as a family business and always operated that way. Over time, Luquette found himself working alongside his cousins—Uncle Lester's boys. Employees who weren't related came to feel a familial bond, too; it wasn't unusual for cashiers, butchers, and other employees to go to work for the supermarket and stay on for twenty years or more.

Years went by and the business grew. In 1984, the family opened a second store in the neighboring town of Delcambre—it rhymes with "welcome"—straddling the Vermilion/Iberia parish line three miles to the east along Louisiana Highway 14. Like the original location in Erath, the Delcambre supermarket was embraced by locals who appreciated the convenience of grocery shopping close to home and, by extension, the validation the store provided to the small community as a place worthy of having its own local institutions. The family opened a convenience store at a busy intersection on the outskirts of Delcambre in 1998, offering the usual array of gasoline, beer, snacks, lottery tickets and—a must for any small store hoping to be successful in Cajun country—hot food.

Luquette's responsibilities grew as well. His uncle, who he always called "Mr. Lester" at the store as a sign of respect, eventually was sufficiently impressed with his hard work, trustworthiness, and business sense to promote him to vice president and general manager, in charge of operations at all three stores.

All was well . . . until September 24, 2005.

Champagne's Supermarket, Erath.

Ricky Luquette, general manager of Champagne's Supermarket.

Luquette lived in Henry, a ten-minute drive south from the Erath store past the cane fields and the sprawling natural gas pipeline junction known as the Henry Hub. As Hurricane Rita approached South Louisiana, he sent his family to bunk with friends in Abbeville. He stayed at home, on alert for emergency situations that might arise at any of the stores, all of which had been shut down in advance of the storm's arrival to allow employees to evacuate or take necessary precautions for themselves, their families, and their homes. Lester had remained at his own home, near the Erath supermarket.

The eye of the storm pushed past the Louisiana coast during the night, well offshore but still strong enough to rattle Vermilion Parish. Luquette's cell phone rang early on that Saturday morning. It was Delcambre's chief of police, James Broussard, with ominous news.

"Rick, water is coming into the store," Broussard told him.

Luquette promptly called his uncle and relayed the report.

"What?" a wide-awake but surprised Champagne replied. "It's still high and dry here in Erath."

Luquette dressed hurriedly, drove over to Erath to pick up Champagne and then sped across to Delcambre, all in a span of twenty minutes. Highway 14 sits higher than the land along either side of it so it was still passable as floodwater from Rita's storm surge began sweeping across the landscape. He stopped his vehicle in front of the Delcambre supermarket, and they got out for a closer look. What they saw was shocking.

"We already had two foot of water inside the Delcambre store. The water was just rushing in," Luquette recalled later. "There was nothing you could do. My heart sank."

Once the storm surge riding Rita's leading edge had reached Vermilion Bay, it pushed up Bayou Carlin and spilled over into downtown Delcambre from the bayouside docks, which for decades had served the town's fleet of shrimping trawlers. For anyone who had not evacuated, there was nothing to do about the rising water except try to get away from it. Rita was on its way to flooding 90 percent of the town.

Their convenience store, several blocks farther away from the bayou, was still dry for the moment. Within a few hours, though, the flooding reached that area, and there was no keeping the water out of that building either.

Luquette and Champagne returned to their flagship store on the street named for the "father of Erath," Dr. Joseph Kibbe, a graduate of Tulane University in New Orleans who established a medical practice, a pharmacy, and a post office in the town once the Southern Pacific Railroad came through in 1893. Dejected, they resigned themselves to the wrath that Mother Nature seemed determine to unleash on southern Vermilion Parish that day.

It wouldn't be the first time. Erath had endured significant flooding before, most notably in 1984, when a storm inundated half the town; the railroad tracks running east-to-west through Erath were sufficiently elevated to serve as a levee, keeping floodwaters out of the southern part of the town where most of the businesses, the high school, and the Catholic church are located. Erath had been battered by hurricanes before, too—Audrey in 1957, Andrew in 1992, and unforgettably, Hilda in 1964. Not otherwise remembered as one of the major hurricanes to strike Louisiana in modern times, Hilda killed eight people in Erath, in one tragic moment.

On the afternoon of October 3, 1964, just hours before Hurricane Hilda made landfall at Marsh Island, on the far side of Vermilion Bay, several volunteers had gathered in the Civil Defense office at City Hall to monitor citizens' band and ham radio transmissions. Outside, the town's water tower was buffeted by increasingly harsh winds as the eye of the storm took dead aim on Vermilion Parish. In one fateful instant, either from a hurricane-force wind burst or a tornado spawned by the storm—no one knows for sure—the legs of the water tower buckled, and the 10,000-gallon water tank crashed onto City Hall, crushing the red brick building.

Five men standing around the front door of the building were pushed away to relative safety by the wall of water from the ruptured tank. Seventeen-year-old Civil Defense volunteer Martial Broussard survived the impact but was trapped by a metal beam and other debris for almost an hour until he could be rescued. Eight others died in the wreckage of the Civil Defense office: Scotty Bernard, nineteen, a student at the University of Southwestern Louisiana; Otto "Cowboy" Bourque, fifty-three, a city policeman; brothers Duffy Broussard, twenty-eight, an appliance store worker, and Vernice Broussard, twenty, a hardware store employee who had been Bernard's classmate in the Erath High School class of 1963; Camile Brown, fifty, a former sheriff's deputy and City Council member;

Felix Dubois, fifty-three, a farmer; Clifton J. Dugas, thirty-three, a construction worker; and Eutis "Noo Noo" Menard, fifty-three, the janitor at Erath High School.

With the community in shock from the tragedy, a single funeral service for all eight victims was held two days later at Our Lady of Lourdes Catholic Church. In the coming months, at the site of the demolished City Hall, the town built a new one, cylindrical in shape to better withstand hurricane-force winds. The building was dedicated to the men who died there during Hurricane Hilda.

Forty-one years later, Ricky Luquette found himself at that spot, staring down the effects of Hurricane Rita, as the counter-clockwise whiplash effect around the center of the massive storm sent an unexpected post-landfall storm surge into lower Vermilion Parish.

Luquette and his son Kyle were members of the Henry Volunteer Fire Department. As the water rose throughout the day across the parish, eventually swamping Erath, they joined in search-and-rescue operations through the town and into surrounding rural areas. In some cases, residents had ignored authorities' warning to evacuate. Others had indeed moved inland but drove back home after the storm passed; they were caught by surprise as the water started rising around them even as the weather was improving. While his son and nephew joined other firefighters and police officers heading out in boats—launched from the lone traffic light at the center of town—to pluck frantic residents from flooding homes and, later, rooftops and trees, Luquette hung back at City Hall, assisting the victims as they were brought to safety.

"We rescued people until dark," he recalled. "After that the National Guard got out of here, the police department got out of here. The only ones that actually stayed were the people that were manning City Hall. I decided I would go ahead and sleep over here in town, instead of going back to meet my wife in Abbeville. I figured I had more to lose, I guess. We had already lost two businesses and I can't afford to lose anything else. You heard of the looting that went on in New Orleans? We didn't think we were going to have that problem in Erath—and we didn't—but I didn't want to take that chance of anything like that happening. We spent the night inside City Hall; we were just a handful."

Throughout the night, calls came in to City Hall from the parish's 911 switchboard about frantic, stranded residents surrounded by floodwaters in the darkness. With no street lights due to the storm-induced power outage and streets covered in water in the dark, Erath authorities—who had tried to rescue those very callers while it was still daylight—stayed put during the night.

"We got all kinds of phone calls throughout the night, from 911, asking us to go rescue people. I called them back," Luquette said. "I said, 'We're not going.' They said, 'You have to.' I said, 'No, we don't. Everybody who's calling you now for us to go rescue, we had somebody there at one of those houses throughout the day. They refused to be rescued then—we're not risking anybody's life now.'"

For small-town firefighters, law enforcement officers, and other first-responders who had lived in the same Vermilion Parish communities all their lives, the Rita experience was unprecedented.

"You've never had anything like this before," Luquette said. "As far as being prepared, nobody was prepared for this. Not even OEP [the parish Office of Emergency Preparedness] was prepared for this. The fire departments, everybody did their share as far as search and rescue." And when the water went down, firefighters in Henry and many other communities found that their fire trucks, firefighting equipment, and in some cases entire fire stations had been destroyed by the storm.

Positioned half a block from City Hall at virtually the highest point in Erath, ten miles northwest of Vermilion Bay and ten feet above sea level, Champagne's Supermarket took on two feet of water, which took two days to recede. At the convenience store in Delcambre, the water only got ten inches high. Inside the Delcambre supermarket, though, the flooding finally topped out at more than seven feet. A CNN report that night on Rita's impact across coastal Louisiana showed footage from the front of the store. The front doors were not visible, completely submerged by the floodwaters.

When the water engulfed the Delcambre store, it actually pushed the building off its footing. Between the flooding and the structural damage, the building was rendered a total loss—a total, uninsured loss. There was flood insurance in place on the Erath supermarket and the convenience store, but Luquette and Champagne had felt they couldn't afford the flood insurance premium on the Delcambre supermarket and so they did without. Adding insult to injury, as Rita was laying siege to the building that fateful day, a fire broke out, engulfing all of the electrical panels and threatening to send the structure up in flames. Had the fire burned the store down, insurance would have covered that. But the rising floodwaters extinguished the flames.

The owners were able to salvage the convenience store and re-open for business there three weeks later. Both supermarkets were ruined, though.

That wasn't the worst of it.

"We were ten family members just on the Champagne side that lost our homes," Luquette said. "Ten family members. Lester's boy lost his. I lost mine. Lester's sister lost hers. He had two nieces that lost theirs. This is all in the Henry-Delcambre-Erath area. My sister lost hers. Two of her kids lost theirs. The biggest part about it—the majority of the family was all working at the store. All of a sudden, everybody's unemployed."

As in nearby Delcambre, about 90 percent of the buildings in Erath flooded. It took two to three days for the flooding from Rita's storm surge to recede. Then began the long, slow process of recovery. It was difficult for individual homeowners, difficult for businesspeople, difficult for entire communities.

These were areas not prone to frequent flooding. Most houses were built on slabs or perched on small concrete pillars just a few feet off the ground. Flood insurance was something that many people felt they did not need. And when Rita proved otherwise, thousands of storm victims throughout the lower reaches of Vermilion Parish—as elsewhere in coastal Louisiana—were left to navigate a maddening, often booby-trapped bureaucratic maze before finding assistance from FEMA and the state's Road Home program, which Gov. Kathleen Blanco established to funnel federal relief payments to homeowners.

One such resident, interviewed twenty months after Rita hit, couldn't spit the words out fast enough as she tried to convey the frustration the recovery process had dealt her as she struggled to rebuild her flooded home and her life. Financial assistance for her losses—house, cattle, and property—was difficult to obtain, slow to arrive, insufficient to cover all the damage and then, adding insult to injury, deemed taxable income: "I don't want to deal with anything. My taxes. The Road Home. The government? *Don't talk to me about the government.* I've had it. I'm tired of jumping through hoops. If you're gonna help me, help me. I ain't asking for the handout—

After Hurricane Rita flooded Delcambre, young men take to the streets in skiffs to help residents needing to evacuate.

Ironic sign in the Hurricane Rita floodwaters at Delcambre High School.

you wanted to give it to me. They gave it to me, and then they sent me 1099s for it. Come on!"

What began as an explanation ended up a tirade, the furious woman yelling not so much at her visitor but at the president, the governor, and legions of anonymous federal and state bureaucrats, all of whom she blamed for the difficulties besetting her and so many others she knew.

As the weeks turned to months, Luquette and his uncle kept thirty of their seventy employees on the payroll—and kept paying their health insurance. Those that weren't family were like family, and most had lost their homes and most of their worldly possessions in the flood. Luquette and Champagne felt an obligation to help them as much as they could, even though the business was generating no income except at the convenience store. In the meantime, they wrangled with insurers and government agencies over the debilitated Erath store's hefty hurricane damage claims, to no avail.

"The insurance didn't want to pay anything," Luquette said. "You contact FEMA to help with your clean-up bills—between the two stores, we had $600,000 worth of clean-up bills. We had to hire hazmat to come in here to clean up. With all the odors and everything else, I didn't want to risk anybody's lives. So we contacted FEMA, I contacted every elected official I could think of, I applied everywhere—nothing. FEMA's response was, quote-unquote, 'We do not help business people.' Who in the hell do they think business people are?

"Six hundred thousand dollar clean-up bill, out of our pockets. We fought the insurance for the longest period of time. We had to end up suing to start collecting some money."

By the summer of 2006, Champagne's was still no closer to opening. Luquette and his uncle were frustrated, but the business had taken such a loss, it could never seem to gain any traction toward reopening. The shuttered supermarket in the middle of Erath was an unfortunate but all-too-fitting symbol for the funk that permeated the community amid sluggish recovery efforts.

Then, in August, Luquette and Champagne were visited by their banker, Mickey Broussard. Ever since the store opened in 1968, it had done business with the Bank of Abbeville, first with Broussard's father Roy, then with him. Broussard met Luquette and

Champagne one morning at the Delcambre convenience store.

His message to them was simple: Champagne's Supermarket needs to get back in business. Erath's recovery is depending on it.

Broussard was well aware that they were still wrangling with insurance companies to try and cover at least some of their losses, and that they were running out of money.

"I know what y'all are doing," Broussard told them. "Come to the bank. Let's get started. We're going to front you guys the money. Don't worry about collateral. We'll take care of everything later. But you guys need to start now because your communities need you."

Until that moment, the owners thought they might never climb out of the rut where Rita had dumped them. Every conversation they had seemed to circle back to the same theme: What are we going to do? Do we want to reopen? Yes, we do, but we're waiting on insurance. The insurance company won't pay our claim. We can't afford to start up again without that insurance money. What are we going to do?

Then, on a handshake, Champagne's turned the corner. In that instant, the burden of worry that Luquette had been shouldering for many long months was lifted.

Luquette didn't see the offer coming, but he should have. Given the family's relationship with its bankers over the years, it should have been no surprise at all. Back in 1968, Champagne applied for a loan at the Bank of Abbeville to get the supermarket started on a secure footing. Roy Broussard studied his business plan, checked out the location and decided to loan him more money than he had requested. Ever since, the folks at Champagne's Supermarket swore by the Bank of Abbeville.

That devotion carried over into their personal lives as well.

Back in 1974—a year after he graduated from Erath High School, and a year before he got married—Luquette decided to buy a little two-bedroom house on the corner of LeBlanc and Hill streets in Erath. The asking price: $6,000. At eighteen years of age, he already had been a Bank of Abbeville customer for six years, having opened his first savings account there at the age of twelve, soon after he went to work for his uncle at the supermarket. He went to the bank and met with Roy Broussard about a mortgage.

"Mr. Roy, I'm getting married next year, and this little house on the corner came up for sale, and I'd like to buy it," Luquette explained.

"How much do they want for it?" Broussard asked.

"Six thousand dollars," the nervous teenager told him.

"I tell you what," the banker replied. "Let's get in my car and go ride to Erath. Come show me the house."

The drive took less than ten minutes, but for Luquette it seemed to last forever. Broussard, who was sixty-two at the time, tried to make small talk with his young customer, and Luquette struggled not to show just how nervous he really was. He always thought of "Mr. Roy" as a nice old man, but at this moment Luquette was petrified.

They arrived in Erath, and Luquette showed him the house. Broussard walked around, inspecting it thoroughly. Luquette, by now even more nervous, stood by quietly.

After checking out the place, Broussard asked him, "How much do they want for it?"

"Six thousand," Luquette reminded him.

The banker shook his head.

"No, I'm not going to loan you six thousand," Broussard told him.

Luquette's heart sank. In that moment, he couldn't even respond as he struggled with the implications of the verdict he had been dreading since before he ever walked into the bank that day.

"No," Broussard continued, "I'm going to loan you $8,000. You need some repairs on the house before you move in."

For the teenage businessman-in-training, it was an important lesson.

"That's the type of people that they are," Luquette would say more than thirty years later as he recalled the exchange. "They're down-home people. You cannot put a value on that."

As his responsibility at the supermarket increased, Luquette saw to it that hometown service was not just a slogan but a way—the only way—to do business. Elderly residents around Erath appreciated that when they could get a ride into town to the pharmacy, they could walk from there to Champagne's and do their grocery shopping, and someone from the store would take them home. Or they could call the store, an employee would pick them up, take them to the store, and bring them back home when they were finished shopping. Or they could just phone in their grocery order to the store and someone would deliver it to their door.

Community connections like that are hard to come by, and the folks in Erath were not likely to take them for granted. It was no surprise, then, that when a team of national planning experts sent from the state capital politely suggested in Rita's aftermath that the entire town be relocated to higher ground to minimize the threat of flooding from future hurricanes, the recommendation didn't go over very well.

Residents took offense at the suggestion by architect and urban planner Andres Duany that the best way to sustain the place they called home was to give up on it and start over somewhere else. The initial public meeting about the post-Rita future of Erath had barely ended when Mayor George Dupuis' phone started ringing off the hook, mostly from older citizens insulted by the notion that Erath as they knew it wasn't worth saving and worried that the proposal might actually come to pass.

Dupuis understood his small-town Cajun constituency in a way that the experts did not. He assured residents that Erath was there to stay, and he wondered if the out-of-town planners were "idiots."

Luquette dismissed the notion as well.

"It's not going to happen," he said. "It's *not* going to happen. Sure, it's so much easier for somebody to come here and tell you what to do with your own life."

Duany's planning team appeared gratified that local citizens were so engaged, and it soon modified its original idea by suggesting instead that Erath expand north of the current town limits, developing an area on higher ground as an option—not a mandate—for flood-threatened residents. Duany envisioned a self-contained site, with shopping, schools, and parks within walking distance of houses. The mayor said he saw merit in the new proposal, and Duany, who had worked on other what's-old-is-new-again New Urbanism projects across the country, began to explore possible sites for what some began calling "new Erath." The effort was stymied, though, by the overabundance of landowners holding rights to tiny slivers of the targeted property—a consequence of Louisiana's forced heirship practice under the Napoleonic Code, whereby children share equally in property passed down from their parents, resulting in family tracts being subdivided into smaller and smaller sections with each succeeding generation. Besides, as often hap-

Erath Mayor George Dupuis.

pens with well-intentioned planning exercises, the public dialogue and published reports failed to identify a substantial and sure-fire revenue source for putting the plans into action, so the recommendation inevitably ended up taking a back seat to more tangible recovery needs.

For Delcambre, Duany's team recommended converting the bayouside, where the town's ever-diminishing shrimping fleet docks, into an upscale, more marketable harbor area. Delcambre Mayor Carol Broussard seized on the idea, recognizing the potential in transforming the workaday dock area into something more appealing to businesses and developers, even at the risk of changing the character of the little town.

The ten-foot storm tide that Rita dumped on the town was devastating, and it drove away an estimated six hundred of the 2,300 people who had lived in Delcambre before the storm. Hurricane Ike hit the town almost as hard three years later, and residents who had stayed in town but failed to raise their homes to meet the flood elevation guidelines set by FEMA were hammered by flooding again. Looking critically upon Delcambre's vulnerability to the elements, Broussard fretted that, absent a dramatic reversal of fortune like that envisioned by the harbor makeover proposal, the little town might just slowly waste away.

Champagne's Supermarket wanted to come back to Delcambre, but it couldn't. In the months after Rita, as the owners wrangled with insurance companies, the damaged store was made available to a Methodist mission and used as a hurricane relief center, warehousing donations of furniture, used appliances, and clothing for the area's needy storm victims. Their court fight against their insurers ultimately proved successful, but by the time the owners saw any money, the Delcambre building—also flooded by Ike—was in such bad shape that it was beyond repair. They often reiterated their commitment to Delcambre, though, and continued to look for alternate sites in the town where they might situate a new store.

"It's killing us not to be in Delcambre right now, but we can't," Luquette lamented.

As the people in Erath, Delcambre, and Henry worked their way back from disaster, Luquette remained acutely aware of the community's struggle. He saw it. He knew it. He was living it himself—not only as a civic-minded citizen active in the Knights of Columbus and the volunteer fire department, not only as the supermarket manager juggling myriad and complex business difficulties arising from three flooded stores, but also on a personal level as a storm victim. He had spent the night after Rita's landfall at City Hall, and his wife, Mary, and daughters, Kimberly and Karissa, had stayed with friends in Abbeville. It was three more days before they could get back to Henry to check on their home.

It's an eerie feeling for an evacuee, waiting for the water to go down, knowing deep down that the damage back home is going to be bad—really bad—but eagerly, almost desperately wanting to see your house for yourself that first time. The Luquettes endured that anxiety like all their neighbors as they waited for the flooding to subside. When they finally made their way into Henry, the displaced houses they saw that had been washed into the cane fields and cow pastures braced them for the worst. Their home, it turned out, was still there, damaged, but salvageable, not entirely ruined or washed away altogether like so many others.

For the next seven months, they lived in the home of their friend Charles "Peanut" Vallot in Abbeville. For five of those months, storm victims Teddy and Gayle Broussard of Erath lived there too.

"Three families in one home in Abbeville," Luquette said. "We were friends, but before we left, we were family. We'd been know-ing each other for years—we see each other, we talk, but you know it doesn't really get personal. But this, before it was all over, our families just knitted together.

"You lose your two businesses, you lose your homes, you lose everything, and you just start over. Everything happens for a rea-son, though. It made truer Christians out of people, it really did. It brought families together that weren't together. It made new fami-lies out of friends."

Back on South Kibbe Street in Erath, Champagne's Supermar-ket finally reopened six months after that fateful meeting with the banker. Freshly painted, shelves fully stocked, new coolers and freezers up and running, the store never looked better. Employees at Champagne's had looked forward to the day for a long time, but they were hardly prepared for what happened when they threw open the doors on January 11, 2007.

The opening day was one big family reunion, filled with hugs, tears, more hugs, a palpable sense of the camaraderie that makes American small-town life so endearing, and finally, shopping carts full of groceries.

"You couldn't imagine it," Luquette said, his eyes welling with tears even months later as he recalled the experience. "Just the warm feeling that everybody had. Everybody became not just friends, but one big family."

And Erath had found its pulse again.

A LEARNING EXPERIENCE

Megan Poole wasn't too worked up about starting high school.

For all that lay ahead—more challenging academics, varsity sports, school dances, club activities—there was no particular anticipation, no anxiety or excitement. It was no big deal.

Such ambivalence was typical for a new high school student in Cameron Parish, where the transition from eighth grade to ninth tends to be less of a milestone than it is for teens in most other Louisiana communities. For students entering ninth grade at Grand Lake High School in August 2005, it meant going to class around the corner—the corner of the old wing of the same school they had attended since kindergarten. There would be the same statue of the hornet—the school mascot—out front, the same cafeteria, the same teachers, and the same few classmates they had always known. As their older brothers, sisters, and cousins had discovered in years past, there wasn't going to be much difference about ninth grade, compared to eighth.

It was much the same at Hackberry High School, population 229 students in pre-kindergarten through twelfth grade, and Johnson Bayou High School, where the entire enrollment from pre-K to twelfth was only 144. The 452 students at South Cameron High School had originated at either of the two elementary feeder schools, but their junior high and high school experience was much like that of their teenage counterparts elsewhere in the parish.

Megan was true to her school, despite its limitations. She didn't begrudge the kids she knew who went to Barbe High School in Lake Charles, with its huge student body and its football team and its broad array of elective courses, but Grand Lake suited her just fine. She was a small-town girl at heart, growing up amid an extended family where her dad, her uncle, and her grandfather all worked together in the community's only roofing business. She was comfortable in this place where she seemed to know everyone, even if it meant not hanging out at the mall and hitting the fast-food restaurants on a regular basis. She was already beginning to appreciate how grown-ups that she knew might work in the city but preferred living and raising their families on this rare patch of solid ground between Cameron Parish's northern boundary and its central waterways and wetlands, away from the noise and the traffic and the other distractions.

Still, in the classrooms along Grand Lake's high school corridor, ninth grade started out as utterly ordinary.

After just two weeks, though, things started to get interesting, at least in world geography class. As news reports about Hurricane Katrina's impact on New Orleans began to emerge, teacher Derrick Robideaux put his students on the case.

"Our teacher got about five or six newspapers, and we each got the front page and had to write a story," Megan recounted. "We each had to find a different angle about Hurricane Katrina, how many deaths there were, how much damage—everyone had to find something and write about it."

The assignments continued for three weeks, and Megan was intrigued by the experience: an event of international significance was happening close to home, and here was a classroom exercise providing students with compelling insights into this history-in-the-making. What's more, the study was prodding them toward concern for real people right here in Louisiana who were living through the ordeal of Hurricane Katrina as it was happening.

"We were all affected by it. For a little while there, we all kind of put ourselves in their shoes. We were very aware," she said. "You never think it would hit so close to home."

She had no idea.

Five weeks into the school year, Hurricane Rita entered the Gulf of Mexico. As the seventeenth named storm of the 2005 hurricane season threatened to impact the southwestern corner of Louisiana, officials in Cameron Parish ordered up the usual precautions. Classes were canceled at parish schools. A mandatory evacuation was announced.

In the coastal communities more prone to storm-surge flooding, those actions engendered a certain intensity. Ninth-grader Tabatha

Harrington and other students at Johnson Bayou High School stacked what they could onto high shelves in their classrooms before leaving the campus. Football players at South Cameron High School were sent home with their uniforms and equipment, with orders to take them along when they evacuated. Stores and gas stations were jammed. At the parish's only traffic light, where Highway 27 meets Highway 82 in Creole, traffic backed up in all directions; some residents were packed up and heading out, while others had to make several trips back and forth to move their cattle and horses to higher ground before evacuating themselves.

To the extent that students were paying attention to Rita, though, many did not anticipate it would be much more than the minor inconvenience that they had come to regard as the typical hurricane experience. After all, Cameron Parish had not taken a serious hit from a hurricane in their lifetime. Most families had packed up in advance of a possible evacuation for Katrina, but when that storm struck the other side of Louisiana on August 29, it had been a regular school day in Cameron Parish, just a little more windy than usual. And Rita was supposed to be heading for Texas.

In the aftermath of the false alarm just three weeks earlier, many students stopped short of hauling out all of their personal belongings as their families prepared to evacuate for Rita. Many just packed a few changes of clothes. South Cameron High School freshman Tony Baccigalopi also grabbed his PlayStation, just to have something to do in the short time he expected to be away from home.

At Hackberry High, algebra teacher Nella Ballou encouraged Jacob Poole, Kelsey Helmer, and their classmates to study during what was expected to be a short break from school, because a big test would be waiting for them once the storm went past.

In Grand Lake, adults and students alike weren't particularly worried about the threat from Rita either. Megan wasn't even aware of the hurricane until she learned that classes were being canceled.

"No one even remotely associated it with a bad connotation," she said. "There was simply no school tomorrow. Even when they called the mandatory evacuation, I don't remember it getting serious until it was almost too late." The Pooles packed a few things and left for a family camp about forty miles north of Grand Lake. More than thity relatives gathered there, safely removed from the threat of coastal flooding. Once the powerful hurricane made land-

Megan Poole, Grand Lake High class of 2009.

The new South Cameron High School, rebuilt after Hurricane Rita. Only the gym remains from the original school building.

fall at the western edge of Cameron Parish, though, it still managed to rattle them with gale-force winds and sporadic twisters.

"The tornadoes were bad up there," Megan said. "I was probably the most scared out of all my family members. I had never been through anything like that before."

Cameron Parish's 1,850 students and their families were scattered throughout Louisiana to await Rita's arrival. Tabatha and her family went from Johnson's Bayou to Merryville, seventy-five miles north of the coast but, like Megan, still in the hurricane's line of fire. Jacob's family went further up the western edge of Louisiana to a lake house at Toledo Bend. The Baccigalopis retreated from the southwestern corner of the state in Creole all the way to Tallulah in the northeast. Kelsey and her family drove from Hackberry up to West Monroe and stayed with friends who were able to accommodate the Helmers' livestock in their barn.

Wherever they ended up, few were prepared for the severity of Rita's impact on what they had left behind. Homes in the coastal communities were smashed apart or washed away. Many of those in the northern part of the parish, in places like Hackberry and Grand Lake, remained intact but were flooded and wind-damaged. Churches were gutted, businesses ruined.

What's more, Rita destroyed four of the parish's six schools—Johnson Bayou High, Cameron Elementary, South Cameron Elementary, and South Cameron High (where only the gymnasium was salvageable)—along with the parish school board's central office on Dewey Street in Cameron. The other two schools, Hackberry and Grand Lake, had substantial damage. In the first few weeks after the hurricane, as the devastated parish remained off-limits to all but emergency workers, evacuees were left to fret from afar about their homes, their jobs, their communities, and their futures. It was at this critical time that Cameron Parish school officials had to figure out how—and whether—to restart the school year, having lost most of its schools and who knew how many of its students. This much was certain: five weeks into their freshman year, members of the class of 2009 had seen the last of a normal high school experience.

The crisis facing Cameron Parish schools in the aftermath of Hurricane Rita was extraordinary. But it was not unprecedented. The school board faced the same quandary after Hurricane Audrey wiped out the parish's coastal towns in 1957. Packing Category 4 winds exceeding 130 mph and a twelve-foot storm surge, Audrey had destroyed not only most of the homes and businesses in and around Cameron but all of the schools as well. As survivors of the killer storm began making their way back—or at least began considering whether to return—the fate of the school system was a major concern for those families with school-age children.

Less than a month after Audrey's landfall, the superintendent of schools notified all teachers and other employees at schools in the coastal communities that the school board would not operate any schools in the flood zone that year, voiding the employment contracts the workers had signed for the 1957-58 school year. The superintendent encouraged the workers to find other jobs, explaining that the destruction left by Audrey made it impractical for the school board to provide any schooling there until schools could be rebuilt.

Once the school officials' action became known, the backlash in and around Cameron was severe. Local leaders were intent on bringing Cameron Parish residents back to revive the communities that had been flooded, and they knew that families would not return without schools that their children could attend. Petitions were circulated, residents began looking on their own for alternative sites where temporary schools could be set up, and board members were pressured to reverse their decision. The board relented. In October that year, an elementary school was opened in makeshift quarters at the VFW hall and the Masonic lodge, and high school students from the area were bused to the school up in Grand Lake.

Nearly half a century later, the school system lost four of its six schools to another vicious hurricane. And in those dark first days after the storm, the devastation was so widespread that no one could know how many residents would be inclined to resettle elsewhere, and how different life in Cameron Parish would become for those who stayed.

Looking out over that depressing landscape, amid the concerns and the confusion, the school board made three specific commitments:

Every employee would have a job.

Classes would be reconfigured with whatever facilities remained useable, so parish schools could retain their respective identities; every student would be taught by his or her own teacher.

All seniors would graduate from their own high school.

To make all that happen, administrators, school board members, teachers, and other employees bought into a simple motto: "Whatever it takes." For Cameron Parish families who returned to find nothing left of their homes but steps and a slab, the revival of the local schools would prove to be not only a relief, but also an inspiration. Clearly, the lesson of Hurricane Audrey had not been forgotten.

Stephanie Rodrigue, the school system's high school curriculum supervisor and personnel director, was put in charge of recovery issues. The first priority was to make immediate repairs to Hackberry and Grand Lake high schools, which were still standing but substantially damaged. As that work got under way, residents began trickling back into the parish to see what was left of the places they had called home.

A mile and a half east of the traffic light in Creole, Tony Baccigalopi's house was gone, period.

"Whenever we got back, it was nothing but house blocks," he said. "There were no walls, no anything. The only thing we ever found from our house was shingles and vinyl siding. The day we first got back there to see it, we were walking through three or four feet of water, waist-deep water, trying to find anything. I remember walking through the water, just keeping your hands down, looking for anything you could find, seeing if you could save it. Dishes, toys, anything."

He would soon discover that almost all of his classmates at South Cameron High had lost their homes, too.

Floodwaters reached eight feet inside Tabatha Harrington's home, on a hill in Johnson's Bayou. When the water receded, nasty black mold soon followed. The family had to leave her two horses behind when it evacuated; one of them was lost in the flood; the other turned up safe, inside someone else's house.

Tabatha was even more shaken by the damage to Johnson Bayou High School.

"The whole front wall was blown out, and everything was pushed out the back wall," she said. "All the wiring was hanging down, and there was just seaweed and mud and junk. Everything was switched around in the rooms, everywhere in the school. Nothing was where it was supposed to be. In the gym, part of the roof was ripped off, and the gym floor looked like a skate park. There

Cameron Parish School Superintendent Stephanie Rodrigue, in the meeting room of the new School Board office in Cameron. Built to replace the old office that was destroyed by Rita, it opened in March 2010.

were big humps everywhere.

"I think I was more upset about my school than our town because we really didn't want to go anywhere else. It was the school where I had gone my whole life. That's where we had always gone. There was never a thought about what happens if I have to go somewhere else. That would never cross our minds, being a freshman."

The Helmers spent two weeks in West Monroe and another week closer to home in Ville Platte before returning to Hackberry. As they drove through the devastated nearby cities of Lake Charles and Sulphur, Kelsey thought to herself, What if we have to change schools? I really don't want to go anywhere else.

Their house was intact, but the floors were caked with mud. The house had flooded about six inches, and that was "just enough to ruin everything," Kelsey said.

Across town, Jacob Poole returned home to find his house had not flooded, but his arrival was greeted by a most incongruous sight.

"When we pulled back into our yard, every cow in the town was in our backyard, drinking out of our swimming pool," he said. "That's the only freshwater they could find."

The flooding extended all the way to Grand Lake, which typically stayed dry even when other Cameron Parish communities were impacted.

"I don't think anyone had an idea of how much it would flood," Megan Poole remembered. "My aunt's house flooded a lot and she didn't bring anything. No furniture was put up—nothing. We had no idea."

Damage totaled $6 million at Hackberry High School and $4 million at Grand Lake High. Sufficient repairs were made in a month's time for faculty and staff to report back there on October 25, just a month and a day after Rita had struck the parish, and classes resumed on October 31.

Platoon schedules were devised. Students from the demolished Cameron schools attended classes at Grand Lake from 7:00 a.m. to 5:00 p.m. on Mondays and Tuesdays and until noon on Wednesdays; Grand Lake students had their classes on Wednesday afternoons, Thursdays, and Fridays. A similar arrangement was made for Hackberry and Johnson Bayou students to alternate using Hackberry High.

The schools had been traditional rivals, so the forced neighborliness was awkward for many students.

"In the beginning, we were rivals with Hackberry," Tabatha said. "We really didn't want to go to school there, but it was better than having to go somewhere else."

The conflict was more pronounced at Grand Lake, where the mid-week transition between the largest of the parish's student bodies would produce "a big hoorah in the parking lot," as one South Cameron student described it.

"It was weird, to say the least, because Grand Lake and South Cameron . . . it's two very different crowds, two very different kinds of people," Megan said, looking back on the situation some time later. "It's just a whole different community. It got a little dicey. They spray-painted our hornet. It got stupid.

"I don't really know anyone from South Cameron, but I think it would have been much harder for them. None of them had homes. They were having to use our facilities. Looking back on it, we probably should have been more hospitable, because they had it worse-off than we did."

By Rodrigue's estimation, in the coastal area from Cameron to Grand Chenier—the attendance zone for Cameron Elementary, South Cameron Elementary, and South Cameron High School—Hurricane Rita destroyed 98 percent of the homes, and the remaining handful was repairable but could not be immediately occupied. The ratio was 80 percent destroyed, 20 percent damaged but repairable in more sparsely populated Johnson's Bayou; 40 percent destroyed, 60 percent repairable in Hackberry; and 25 percent destroyed, 75 percent repairable in Grand Lake.

She explained, "All this meant all of our kids and our employees that were residents of Cameron Parish had these options:

"FEMA trailers, which were very slow coming here, and understandably so because we were all in the same queue behind the people for Katrina. A month later they were still working on Katrina, so we went to the back of the line.

"Lots of moving in with family. That's probably the number one thing that people did.

"Living on property north of here. A lot of people had property to move cows to, in different seasons. If they didn't have a place on that property, they bought campers and stuff—personally bought them and put them on property they owned.

"We had people that stayed in hotels for awhile.

"A lot of people just bought property elsewhere. Houses, not just property—or they bought property and put trailers on them."

Seldom does Mother Nature deal any one community such a drastic blow. For residents old enough to remember Hurricane Audrey, it was the second such calamity in their lifetimes.

"We evacuate, you know. We're used to that," Rodrigue observed. "For a time, we could count on evacuating once or twice a year. And then we had some good years. But every time you leave, you really don't think in terms of not coming back."

About 1,500 far-flung students showed up for classes the first week that the schools reopened, representing 80 percent of the school system's pre-hurricane enrollment. At that point, virtually all of them were still living outside the parish, so buses were dispatched extraordinary distances to get them to school and back. Families remained scattered all across central and south Louisiana, and the school system provided bus service within a seventy-five-mile radius, as far as DeRidder and Kinder and almost to Lafayette.

"We bused as far out-of-parish as we needed to," Rodrigue said. "We had families that were staying in Baton Rouge, and they brought their kids to Rayne. That's the farthest we could go and get back to school at a reasonable time."

After a month of upheaval, students and teachers alike welcomed the semblance of a school routine, even one that only lasted two and a half days a week. Not unexpectedly, though, the usual focus on academics was difficult to sustain.

"We saw a lot of emotional distress," Rodrigue said. "Kids were struggling. Mainly, they were distraught because their parents were upset. There was a lot of that. Little things, like a little one who doesn't have any toys left. They just don't understand that you leave to go on a trip, and you come back and everything's gone."

Counselors were brought in to help students cope. Members of the faculty, meanwhile, found themselves working all the harder to keep troubled or at least distracted students on task while juggling their own conflicts as flooded-out and displaced homeowners themselves.

"You had the parents at home that were frazzled. Then you got to school and we had teachers that were frazzled," Rodrigue said.

School administrators made it as easy as possible for teachers and other school employees to deal with those challenges.

"We pretty much gave them the authority to go meet your FEMA rep, go meet your insurance rep," Rodrigue said. "If you need to leave school and run down there, go do it. That's all we could do. We had to take all our firewalls off the computer, so people could correspond with their insurance company. They could type their list of items they lost and send it on, even though it was hitting up against our no-shopping policy. We relaxed all of that. We encouraged them. Everybody covered for each other. If somebody had to go check on their insurance, they went and checked on their insurance and the person next door watched their class."

The learning conditions were something less than optimal in the first place, and then students were going from a teacher who was under stress at school to parents who were under stress at home. Student performance couldn't help but be affected.

"We didn't lose that much ground, but we lost some ground academically," Rodrigue said. "I think it's a small price to pay for our playing a role in making it as easy as possible for those who want to come home, to come home, and those that want to be educated here, to be educated here.

"A lot of people say they came back because the schools came back. If that's true, I'm happy."

In the spring, portable buildings for the Johnson Bayou students were donated to the school system by a company with industrial operations at Johnson's Bayou. Very few people had moved back there yet, though, and the campus remained too littered with storm debris for the portables to be installed there, so they were set up at the Hackberry campus. From that point, the Johnson Bayou students were able to consider the temporary facilities as their own school-away-from-school.

Grand Lake High continued to host the coastal Cameron students for the remainder of the school year. The hurricane cost the South Cameron High football team its stadium and the heart of its season, but after missing a month the Tarpons were able to regroup and finish the year with a big assist from Lake Charles' Barbe High School, nineteen miles up the road from Grand Lake. Once the coaching staff was able to field a team again, Barbe provided its stadium and a locker room for the Tarpons to play their final three games of the season. Come spring, the Lady Tarpons softball team put together a magical run of nineteen wins and only four losses. Although still displaced with the rest of their schoolmates, team members fought their way to the title game where they defeated District 5-1A rival Merryville 5-0 and brought home the school's first state championship. Weeks later, the Louisiana Legislature adopted a resolution congratulating the Lady Tarpons on their achievement, citing their "strength of spirit" in overcoming "great personal adversity" in the wake of Hurricane Rita.

Throughout that difficult school year, Rodrigue was consumed by the responsibility of steering the school system toward a solid recovery. Just dealing with FEMA seemed like a full-time job. So did working with contractors and insurance companies. And so did monitoring the activities in the two schools that the school system had available to use, and the tear-downs and clean-ups at the four schools that were out of commission. As a school system supervisor, she was determined to formulate a workable plan for making necessary repairs and accommodations so the school year could be salvaged, while safeguarding the school board's budget to the extent possible. She also was burdened by the exacting demands of the FEMA bureaucracy at such a complex administrative level.

Amid all that, she retained a sharp appreciation for what students and their parents were going through, because she was living it, too.

"We returned to, literally, two steps and a slab," Rodrigue said, describing what was left of her home three miles east of South Cameron High School. "We didn't have to deal with debris. There was nothing, nothing left where we lived. My husband had to tell me, this is where we lived. There were no landmarks, because the neighbors' [home site] was just crumbled brick."

Immersing herself in school business, Rodrigue couldn't find the time to make personal arrangements for herself and her family with relief agencies.

"Most people had more time to figure out what they were going to do, but I was busy figuring out what the kids were going to do," she said. "We lived in the living quarters of our horse trailer for six months, because I couldn't take a breath long enough to follow up on a FEMA trailer."

Finally, some friends loaned her and her husband a modest mobile home that their grandson had used when he was in college.

"It's several years old, single-wide, but it was something," Rodrigue said. "It was better than the living quarters of a horse trailer. You went on—you just did what you had to do."

May finally arrived, and the school board kept its word that high school seniors would graduate "at home." At Johnson Bayou High School, that meant setting up a huge tent in the driveway in front of the ruined school; the South Cameron seniors went back to their own campus, too, and graduated on the track at their battered football stadium. Commencement ceremonies were imbued with a greater sense of emotion and relief than usual, serving to recognize not just the achievements of the four groups of graduating seniors but also the perseverance of their communities and Cameron Parish overall.

Over the summer, the portable classrooms for the Johnson Bayou students were moved from Hackberry down to their old school grounds. Even though it would be another year before their school

Valedictorian Katie Young hugs her sister Kandace at the 2006 graduation ceremony held at the remains of Johnson Bayou High School, which was destroyed by Hurricane Rita.

building would be rebuilt, just being back on their familiar campus was meaningful for the students.

Rodrigue arranged for the school board to acquire forty-four portable classrooms to accommodate the Cameron and South Cameron students. Those temporary buildings were set up in four "pods" on the South Cameron High School campus during the summer. They weren't cheap, but Rodrigue knew—and the board agreed—that it was important to bring the Cameron-area students back to their home territory, and with conditions at all three school sites in that area ranging from substantial damage to complete destruction, the use of portable buildings was the only short-term option for getting the students out of Grand Lake and back to the front ridge.

"Originally we had a commitment from FEMA for a 'turn-key' temporary facility that we could walk into in January '06," she said.

"It didn't happen in January or in March, so we said we were going to use some local funds and just do it ourselves, and that's what we did. We had to do something."

FEMA later agreed to cover the cost of those temporary buildings.

The portables were set up in the South Cameron High parking lot during the summer, but the occupancy permit wasn't issued until the day before the 2006-07 school year started. That final day of summer vacation was a frenzy of activity at the school site, as teachers, parents, and other volunteers hastily moved new desks and other equipment and supplies into the classrooms to get them ready for the start of class the next day.

The school board also launched plans to build a new South Cameron High School there, elevated to guard against future storm surges. The ground-level gym was repaired, along with the football stadium, but everything else about the old school was lost and had to be rebuilt to more stringent specifications—higher and stronger—for new construction in the Cameron floodplain. Before long, the board officially consolidated the three schools—Cameron Elementary, South Cameron Elementary, and South Cameron High—permanently merging the classes from pre-kindergarten to twelfth grade into a single facility, as happened by necessity since Rita and had long been done elsewhere in the parish.

Not long into the school year, students at all four parish schools were assigned to write a story or draw a picture about how Hurricane Rita affected them and their families. The students' personal accounts, along with other keepsakes from parish residents, were placed in a time capsule that parish officials buried on the courthouse lawn to be unearthed in fifty years. As time went on, the school year proved to be decidedly less chaotic than the one that had preceded it, but Rita's lingering impact was never far below the surface.

The South Cameron football team played its home games at Barbe in Lake Charles again for the 2006 season. With the students attending class back on their own campus, the team was able to practice at a field nearby in Creole. Although convenient, that also proved to be perilous.

"Every day somebody would be getting cut up because our practice field had so much debris—aluminum door frames, glass, metal, tin," Tony said. "Every day somebody was getting their arms or their legs gashed open from some kind of debris on the practice field. Finally, one day we didn't even practice. We just went out there and grabbed trash off the field so we could practice without somebody getting hurt from it."

Fourth-graders learned about hurricanes and storm erosion in their science classes. While elementary students elsewhere were making clay-model volcanoes for the standard vinegar-and-baking-soda eruption experiments, the Cameron students built replica sand beaches inside water-tight containers, reinforced them with a variety of techniques and watched how their efforts held up to "wave action" when they added water and sloshed it around. It wasn't hard to interest students in the lesson, South Cameron teacher Peggy Griffith saw; they understood the concept from first-hand experience.

Before long, there would be more first-hand experiences, more hard lessons.

Hurricane Ike took aim at Cameron Parish just weeks into the 2008-09 school year. The students who had barely started their freshmen year of high school when Rita hit were now just weeks into their senior year, and all they could think was: here we go again.

Tony and his dad had been back in Creole for about a year, living in a trailer just long enough for him to graduate from South Cameron High.

"We had kind of a false alarm with Gustav on Labor Day weekend, " he recalled. "Everybody was worried about it coming here, and then it ended up turning. For Ike [the next week], I think I knew in the back of my mind it was going to happen all over again."

He was right. After evacuating for the storm's landfall, father and son returned in a few days to find their mobile home gutted.

Throughout Cameron Parish, there would be more flooded homes and churches, more damaged schools, more displaced families, another wave of disruption for youngsters who already had endured more challenges than most people face in a lifetime.

By this time, the school board had promoted Rodrigue to superintendent of schools. Based on the Rita experience, she was well-positioned to lead the school system's recovery from Ike.

The Johnson Bayou and South Cameron schools were clobbered again, so those students found themselves on the move once more. Johnson Bayou students returned to their familiar exile at Hackberry High until the Christmas break. Instead of sending South Cameron students back to Grand Lake, though, the school board found them an unorthodox, if temporary, home. For the rest of the fall semester, an old bingo hall in Lake Charles was pressed into duty as an ersatz South Cameron High School, its vast interior partitioned to create classrooms while business went on as usual at Walmart and Lowe's next door.

Come January, students were back on their home campuses, and although conditions might have been unsettling, they and their teachers forged ahead with the rest of the school year. As graduation approached, members of the Class of 2009 had mixed emotions about the way in which Rita and Ike had bookended their high school experiences.

Tabatha Harrington was quick to credit her Johnson's Bayou school community with keeping her grounded throughout the traumatic four years, which also included the death of a classmate in a traffic accident on the morning of junior prom as he was en route to pick up his tuxedo.

"I think being in school with our friends kept me more sane than anything," she said. "Seeing the faces that I'm used to seeing all the time, even though we weren't in our own school, helped me get through it, because I knew I wasn't the only one going through it."

In Hackberry, where the flooding was much worse for Ike than it had been for Rita, Kelsey Helmer was struck by how many people she saw move away from her small town rather than stay and rebuild like her family did. Meanwhile, the disasters convinced her classmate Jacob Poole not to return to his hometown after college.

"I don't want my kids to go through what we went through," he said. "I won't live anywhere near the coast.

"It's nice to live in a small community with everyone you know, but . . ."

He shook his head, leaving the sentence unfinished.

Tony Baccigalopi was looking forward to graduating from South Cameron High. After losing two homes in three years, seeing his parents divorce, and enduring the repeated unheavals that major hurricanes caused coastal Cameron Parish, he was eager to move on and make a fresh start.

And Megan Poole, wise beyond her years, recognized that the peculiar circumstances of her time in the senior wing at Grand Lake imparted a lesson that she never anticipated from a high school curriculum.

"For me, you realize that there are some things that you'll never be able to grasp through human reason alone," she said. "Now I know on a personal level that a natural disaster is one of those."

She shared the anguish of her aunt and uncle and numerous classmates whose houses flooded in both hurricanes. She saw tremendous damage inflicted on her beloved school. She looked beyond a superficial rivalry to view with compassion the displaced students who had landed in Grand Lake after losing not only their own school but, in most cases, their homes as well. She came to realize that the experiences made her not only a stronger person, but a better one, too.

"You grow in areas like sympathy," she said. "You grow in areas like endurance, patience, on a spiritual level and as humans. You realize there's something so much higher than you to get through it."

DO NOT HARM
MY CHILDREN

For the Most Reverend Glen John Provost, it was impossible to overstate the impact of the Catholic church on Louisiana's Cajun country. Over many years, in lectures and pastoral letters, he emphasized that the Catholic faith is why the Cajuns exist in Louisiana in the first place.

"They are here because of that," he would say. "It's really very much the *raison d'être.*"

Their Acadian ancestors were deported from Nova Scotia in the mid-1700s for refusing to take an oath of allegiance to the British crown, and that refusal hinged largely on the de facto insistence that they renounce their Catholicism and embrace the Church of England. Acadian exiles arriving in South Louisiana found Catholicism already deeply rooted among the Spanish, German, and French settlers who had preceded them there; the first Catholic church in the region was established on the "German Coast" along the Mississippi River at Destrehan in 1723. Almost three centuries later, Catholic influences still run deeply through life in Cajun Louisiana, and that faith would prove to be a key to the recovery from Hurricanes Rita and Ike for individuals and communities alike.

Possessing a master's degree in English literature and conversant in French, Spanish, and Italian, Provost was a student of history—the church's, which had interested him since childhood; his own, going back to Provost and Blanchet ancestors who had come directly from France to settle in South Louisiana two centuries earlier; and that of the Acadians, whom he regarded as "tenacious, hard-working, hospitable, and rooted in the land." Provost was a parish priest at Our Lady of Fatima Catholic Church in his hometown of Lafayette when, in the summer of 2005, he celebrated the thirtieth anniversary of his ordination at St. Peter's Basilica in Vatican City. In the weeks that followed, though, he would see how past experiences could not have prepared him for the ordeals wrought by the ravages of that hurricane season and the impact they would have on himself, his church parish, and his homeland.

It started at the end of August with Hurricane Katrina, which skirted the eastern edge of Acadiana and targeted the New Orleans area and the Mississippi Gulf Coast.

"The first reaction that I had, and I think it was the reaction of everyone else who had lived in Louisiana all their lives, was, 'We'll be back in two or three days and we'll get back to normal and start picking up the pieces and cleaning up,'" Provost recalled. "But as we watched the episode unfold, you became aware that this was a storm like no other. We've all seen them do strange things—turn course as they reach the coast, dissipate as they reach the coast, go inland and remain inland and come back out toward the Gulf. But this, when the levees broke, I think my reaction was shock. Then I realized, I think within the first day I realized, that this was a catastrophe of mammoth proportions and that we would be having an influx of displaced persons as well as refugees from the storm into Lafayette, of every sort, and that we needed to accommodate them in some way."

At Our Lady of Fatima, a church of about 1,300 families and 4,000 total parishioners—medium-sized by Lafayette standards—Provost and members of his parish staff quickly organized a variety of relief efforts for many of the thousands of refugees who poured into Lafayette once New Orleans flooded. Headed by the parish's social services director, Virginia Webb, the relief committee met regularly—several times a week at first. It soon recognized that the needs were changing daily, if not hourly, in those first chaotic weeks after Katrina.

The church owned a rental property; it provided that home to a displaced family. The Lafayette Parish Sheriff's Office needed to house a National Guard unit from Washington, D.C.; the committee offered the Fatima parish hall. An army of volunteers had assembled at the University of Louisiana at Lafayette's Cajundome arena to feed and care for thousands of storm victims who had landed there, but no one had arranged to feed the feeders; the committee recruited church members to start cooking for the volunteers.

"The list went on and on and on," Provost said.

Like the even larger city of Baton Rouge to the east, Lafayette was overwhelmed by a crush of Katrina refugees, suddenly homeless, devoid of possessions, and bereft of hope. Some were trying to get some footing for themselves and their families; others merely seemed dazed and confused in their new surroundings.

"It was just incredible—I'd never experienced anything like that in my life," Provost said. "You'd walk the streets of Lafayette or you'd be in your car and you'd be stopped by somebody displaced and they needed something. They needed food, they needed clothing, they needed lodging, whatever it was."

Like many churches across South Louisiana, Our Lady of Fatima collected donations for Katrina victims, and the church relief committee determined that one effective means of assisting the needy was to provide retail gift cards.

"We would go to Walmart, get Walmart cards and keep them in our pockets and just give them out," Provost said. "You'd go to a restaurant and the owner of the restaurant would tell you, 'I just hired a waiter and a cook and a busboy. They were displaced by the storm and I felt sorry for them. I had to offer them work.' I'd say, 'Here, give them the cards.' I had three nuns flag me down in a parking lot, screaming after me, 'Father, Father, Father,' and they came running up to me. They had been displaced and they were living in a farmhouse outside of Carencro. Could I please tell them where they could get gift cards? I had some in my pocket."

For the first three weeks of September, the church focused its efforts on ministering to Katrina victims: sheltering the homeless, feeding volunteers who staffed refugee centers, finding new schools for displaced students, and arranging wakes and funerals for those who died during the evacuation. Many parishioners had relatives or friends from New Orleans move in with them temporarily. It was hard to find a member of Fatima parish—or most other churches in Lafayette—who didn't have a relative or an acquaintance affected by the calamity.

Then the region was slammed by Hurricane Rita, and the impacts were felt much closer to home. Many churches in the lower reaches of the Diocese of Lafayette—St. Anne in Cow Island, St. James in Esther, St. John in Henry, and others—were flooded and badly damaged by Rita's storm surge, just like the homes of most

Bishop Glen J. Provost.

of their parishioners. The extensive flood damage throughout the region also kept evacuated church members dispersed, making it all the harder for battered churches to regain their footing.

In Lafayette, parishioners who once again dodged direct storm impact immersed themselves in relief efforts for others. Shelter and meal services for evacuees were expanded to care for Rita victims as well as those from Katrina who were still in the city. The Catholic Services Center provided financial help for storm victims with prescription medications, rental deposits and payments, utility bills, and transportation needs; another diocesan outreach program sent workers to the homes of elderly flooding victims to treat the houses for mold. The predominantly African-American congregation of Our Lady Queen of Peace collected school uniforms, backpacks, and school supplies for children of the storm. St. Elizabeth Seton parishioners established a fund to help local residents who had taken in hurricane victims.

Over on Johnston Street, the Fatima relief committee turned its attention to the immediate, specific needs of neighbors from nearby Vermilion Parish. Church members collected and delivered bedding, kitchenware, cleaning supplies, nonperishable food, and over-the-counter medications for people from Delcambre, Erath, and elsewhere who lost most of their earthly possessions to the flood.

The opportunity to provide aid and comfort in the aftermath of Katrina and Rita increased parishioners' awareness of the needs of the less fortunate and, according to Provost, reinforced for them the Catholic tenets known as the evangelical counsels—specifically, the need to provide for others in a spirit of compassion.

"That's what was going on—it was the corporal works of mercy," Provost said. "It was burying the dead. It was visiting the sick. It was feeding the hungry. It was all of that. It gave a name to all of these works that were being done and that were necessary and immediate."

Just to the west, in the Diocese of Lake Charles, the challenges were even greater. The roof of the diocese headquarters in downtown Lake Charles collapsed during the storm, and all thirty-eight churches in the diocese suffered some damage from wind or water—the closer to the Gulf, the more dire the circumstances. Throughout the communities near the coast, Rita's devastating storm surge reduced homes to rubble or washed them away in their entirety,

Two days before Rita's landfall, NeNe's Kitchen asked for divine assistance in downtown Cameron.

and the churches didn't fare much better. Parish churches such as St. Eugene in Grand Chenier, Our Lady Star of the Sea in Cameron, and Sacred Heart in Creole—Cameron Parish's original Catholic church parish, founded in 1890—were inundated and badly damaged but deemed repairable. Three smaller mission chapels—Immaculate Conception in Grand Chenier, St. Rose of Lima in Creole, and Holy Trinity in Holly Beach—were completely consumed by the flood, with no remnants of the buildings left behind except concrete slabs.

For many residents, it would be months, not weeks, before they could set about rebuilding their homes—if they chose to return at all. Many remained in evacuation mode, bunking in with relatives or friends or living in mobile homes or travel trailers in Lake Charles, Grand Lake, and other communities. FEMA trailers were hard to come by and slow to arrive. Many storm victims who were entitled to them were so frustrated by the bureaucratic entanglements and delays involved in acquiring FEMA trailers that they gave up and bought their own campers for temporary lodging.

The Rev. Joseph McGrath, pastor of Sacred Heart Catholic Church, did that; so did Susan Johnson, the church's administrative assistant, and her husband, Tony, and members of their extended family. McGrath and the Johnson family grouped five travel trailers together on a single piece of property far inland from Creole; while they tended to the torturously slow business of repairs to their homes and, in McGrath's case, the church and rectory, they lived together in their little camper settlement which they dubbed "Johnsonville."

Any hope of a quick recovery in lower Cameron Parish was dashed by a broad range of obstacles. Insurance companies disputed many homeowners's claims for hurricane damage. Financial assistance promised by FEMA, the state's Road Home program, and other government agencies set up to assist storm victims was tardy and insufficient. Building materials and workers were in short supply. Government requirements that some new homes be built to higher standards than the ones they were replacing resulted in a substantial increase in cost. Consequently, residents remained scattered well into the next year, wanting to return home and make their neighborhoods whole again but finding it difficult to do so.

Throughout that trying period, local church communities were

Father Joseph McGrath, right, inspects flood damage to Sacred Heart Catholic Church in Creole following Hurricane Rita.

Hurricane Rita destroyed the religous education building behind Our Lady Star of the Sea Catholic Church in Cameron.

a vital touchstone for this overwhelmingly Catholic population. The church buildings themselves might have been in shambles, but a church is always more than brick and mortar and stained glass windows. In the midst of such upheaval, there was inspiration to be found in the practice of the faith and comfort to be taken in the fellowship that derived from gathering for Sunday Mass.

The diocese was without a bishop at that time. In mid-March 2005, just three weeks before his death, Pope John Paul II had transferred Bishop Edward K. Braxton from Lake Charles to a diocese in southern Illinois, without designating a replacement. Pope Benedict XVI was installed at the Vatican on April 19, but when Hurricane Rita struck five months later, the bishop's seat in Lake Charles was still vacant. In the interim, Monsignor Harry Greig handled administrative functions for the diocese in addition to his duties as pastor of St. Mary of the Lake Church in Big Lake, in the inland northern section of Cameron Parish.

While coordinating a diocese-wide response to the disaster and dealing with severe flooding that Rita's storm surge had inflicted on his own church, Greig arranged for a regular Mass time every Sunday morning at his parish's mission chapel specifically for the congregations of the damaged churches in Creole, Cameron, and Grand Chenier.

"That was important," Greig said. "It gave the community just an opportunity to be together."

That 8:30 a.m. Mass at St. Patrick Chapel at Sweet Lake soon became an important part of the Sunday routine for parishioners of Sacred Heart, Our Lady Star of the Sea, and St. Eugene who had been displaced across the Grand Lake area, Calcasieu Parish, and beyond.

"One of the most emotional things was just to be at Mass and realize that nobody in that room had anything. Nothing," said school system official Stephanie Rodrigue, a Sacred Heart parishioner.

In Grand Chenier, the pastor at St. Eugene, the Rev. Vincent Vadakkedath, began saying occasional Sunday Masses in the open-air ruins of the church, which still had a front and back but had lost its two side walls. Displaced residents drove in from far and wide to spend those Sundays with their friends and former neighbors.

"That was an emotional time, but it helped everybody," one St. Eugene parishioner remembered. "You'd laugh, then you would cry, then you would laugh again, and when it was time to leave no one wanted to leave. We would have Mass at ten in the morning, and we'd have dinner there. Everyone would bring something and we'd all have dinner there. We'd end up staying until three or four o'clock in the afternoon and no one wanted to leave."

Greig remembered Hurricane Audrey as a boy growing up in Lafayette. He had heard all the stories, but it wasn't until he arrived in the Lake Charles area as a priest that he understood the extent of Audrey's deadly impact on the people and communities that found themselves in its path. With that frame of reference, he would marvel at the determination of parishioners to battle back from the setbacks that Rita handed them.

"It's their rootedness. It's their life," he said. "In smaller communities, you have more of that bondedness. One of the things people missed during the recovery time is being able to know where everybody was and see everybody and be with everybody. There were times for that but it wasn't like what it had been. Throughout it all, though, I found people very hope-filled and faith-filled. It was amazing, their perseverance and their strength of faith and their moving forward."

Inevitably, residents returning to south Cameron fretted over the future of their church parishes. They tried to persuade diocesan officials that restoring all of the storm-damaged churches would accelerate recovery in the region and induce more displaced residents to come back. Grieg, as caretaker of the diocese's administration, had a different view of the chicken-and-egg conundrum.

More than a year after Rita, he offered this perspective: "It's been a slow process, yet at the same time it's been one of combining, trying to minister to people and their needs, while at the same time be realistic in terms of long-term plans. We didn't want to build everything too quickly because if there's no people down there to support that, then what do you do with the buildings? It's a touchy thing, because you're dealing with people's emotions, and of course they wanted everything built yesterday.

"It's not that you don't care, it's that there's a bigger picture you have to consider."

That picture was that some of Lake Charles' largest church parishes had 2,500 Catholic families and only one priest to minister to them, while the churches in Cameron, Creole, and Grand Chenier had less than one thousand families combined. In the aftermath of catastrophe, when recovery was uncertain and everyone was expected to do more with less, it was difficult to commit right away to making those three smaller, far-flung church parishes whole again, each with its own priest.

Mission chapels that had been lost to the storm were not replaced. The matter of reappointment of clergy remained an open question. Once FEMA rated the Cameron, Creole, and Grand Chenier churches repairable, though, plans for rebuilding were put in place.

The task of restoring their church after a storm was nothing new to the faithful of the town of Cameron. Its first Catholic church was St. Joseph, a mission chapel of Sacred Heart parish. Built with donations solicited by Sacred Heart's founding pastor, the Rev. John Engberink—including fifty dollars contributed by Mother Katharine Drexel, the American missionary nun who would be canonized in 2000—St. Joseph Chapel was dedicated in 1894, only to be destroyed by a hurricane in 1909 and abandoned. Another chapel was built in 1914 and named for St. James, but it was lost to another hurricane in 1918. After paved roads reached the isolated settlement in the 1930s, a new chapel was built in 1937 and given the name St. John the Evangelist.

When Hurricane Audrey made a direct hit on Cameron in 1957, the church was badly damaged. Bishop Maurice Schexnayder championed the cause of Cameron's recovery from that deadly hurricane, visiting the disaster zone many times. In 1958, he declared the town's church an independent parish and renamed it Our Lady Star of the Sea. Within a few years, a new church would take shape there, graced with a seven-foot tall statue of white marble commissioned by Schexnayder and sculpted by an artisan of the Tavarelli Marble Co. in Carrara, Italy. Installed on the church's front lawn, the statue depicts the Blessed Virgin Mary, facing down the Gulf of Mexico with her arm around a girl at her side. The inscription reads: "Do not harm my children."

The statue survived the wrath of Hurricane Rita, but the storm blew out windows and walls of the church, washing away most of its contents and damaging what remained. Extensive repairs were undertaken, aided by an unlikely source: the Benedictine Sisters of Erie, Pennsylvania. In the summer of 2006, the nuns at the Mount

Lake Charles artist Elton Louviere painted this mural behind the altar at St. Eugene Catholic Church in Grand Chenier.

This statue of St. Eugene was lost when Hurricane Rita's storm surge struck the Catholic church in Grand Chenier. The statue was found miles away in the town of Lake Arthur.

St. Benedict Monastery began major renovations to their chapel, including leveling a two-foot slope in the floor from the back of the chapel to the front which was difficult for elderly members of the congregation to navigate with their walkers. After learning about Rita's impact on Our Lady Star of the Sea and discovering that the Cameron church was very similar in design to their own chapel, the nuns arranged to donate to the Louisiana church parish 102 wooden pews, five ceiling fans, and dozens of light fixtures. An Erie-based moving and storage company offered the Benedictines a reduced rate on packing and shipping, and soon the pews, fans, and lights were transported from the shores of Lake Erie to serve another faith community on the Gulf of Mexico.

Early in 2007, word arrived from the Vatican that Pope Benedict XVI had settled on someone to fill the long-vacant bishop's post for the Diocese of Lake Charles. The pontiff's choice: Monsignor Glen John Provost of Lafayette.

At a Holy Thursday prayer service for Our Lady of Fatima Catholic School students, Lafayette Bishop Michael Jarrell predicted that Provost, as bishop of the southwestern Louisiana diocese that had been carved out of the Diocese of Lafayette in 1980, would "make sure our western borders are secure." Fatima parishioners bade farewell to their beloved pastor on Easter Sunday, and he arrived in Lake Charles the next day to become the spiritual leader for thousands of Rita victims in Cameron and Calcasieu parishes who were still struggling to rebuild their homes, their churches, their communities, and their lives.

"I am returning to the familiar and friendly city of my first assignment"—as associate pastor of Our Lady Queen of Heaven Church in Lake Charles immediately following his ordination in 1975—"and to a people who exhibited such courage and perseverance in facing the full force of a destructive storm," he said in an initial statement to the seventy-eight thousand Catholics of the Diocese of Lake Charles prior to his installation as bishop there. "These steadfast and strong men and women are an example to us all."

Being introduced to the people of southwestern Louisiana in the midst of the recovery, Provost had to get to know them on the fly, but he understood them from the get-go.

"This is where we live," Provost said. "This is home. There is a rootedness here we have to acknowledge. Our people here have a

Sacred Heart in Creole, Ike cleanup.

Hurricane Ike damage at Our Lady Star of the Sea Catholic Church in Cameron.

history here that's very rich."

Two months after his arrival, he traveled to Grand Chenier to rededicate St. Eugene, the first of south Cameron's Catholic churches for which repairs of Hurricane Rita damage were completed. During the ceremony, the bishop walked up and down the aisles, sprinking holy water on the parishioners and walls of the church. In his homily a short time later, he noted the irony of how the church was so badly damaged by water when the hurricane came ashore but then blessed with water as it was returned to the Lord's service.

"How tame that water was," the bishop said of the ritual blessing. "It wasn't tame almost two years ago with Hurricane Rita—nothing tame about that."

Provost was with the people of the Lake Charles diocese from the outset when Ike came through in 2008, in some instances ruining the repairs to homes, churches, and other buildings that had been two and three years in the making, leaving church members to start over yet again. Sacred Heart Church in Creole was one week away from reopening for the first time since Hurricane Rita when Ike pushed through and ruined all the new pews, never sat upon, and fixtures, never used. The church in nearby Cameron likewise was flooded again when a wall collapsed. The solid oak pews from the Pennsylvania convent were destroyed, along with most of the other contents of the church. Indeed, for all of the Catholic churches in Cameron Parish, Ike was virtually an instant replay of Rita.

As beleaguered members of his flock looked to the bishop for guidance, Provost would marvel at their faith and their courage to rebuild once more, recognizing that they still prayed the prayer that Bishop Schexnayder had composed after Hurricane Audrey. That "Prayer for Safety in Hurricane Season" says in part: "We live in the shadow of a danger over which we have no control: the Gulf, like a provoked and angry giant, can awake from its seeming lethargy, overstep its conventional boundaries, invade our land, and spread chaos and disaster." It goes on to ask the intercession of Mary Star of the Sea, "so that spared from the calamities common to this area," the faithful might "reach the heavenly Jerusalem where a stormless eternity awaits us."

For the second time in three years, the diocese faced serious questions about how to allocate resources to recover from a serious hurricane. Many of the same issues about rebuilding and staffing churches and providing other church-related services followed Ike, just as they had followed Rita. This time, the buck stopped with Provost for those tough decisions.

Through the crisis, the bishop with the historical bent would ponder time and again, with confidence, that the church, as an institution and as a collection of dedicated individuals, would see the people of South Louisiana through it all.

"The church has been and will always be around here," Provost said, when asked about the diocese's role in the region's recovery. "It's not transitory. It's here, its roots are here. There is a permanence to it. There is a longevity to it. Whether you're talking about the yellow fever epidemics in New Orleans in the nineteenth century—when the priests and the nuns of the church stayed during those times of cholera and yellow fever, when everyone else fled—or whether you're talking about the hurricanes, this story of the Catholic church's permanent assistance and remaining with it is there."

DO UNTO OTHERS

On her forty-ninth birthday, Margaret Hebert parked in front of the television in her Lafayette home and started crying.

It was August 29, 2005—the day Hurricane Katrina came ashore. From the comfort of her favorite chair in her dry, air-conditioned, nicely furnished brick home in the heart of Acadiana, Hebert sat sobbing, transfixed as the disaster that would ultimately claim 1,836 lives across four states and scatter hundreds of thousands of refugees across the country began to play out before an international television audience. First the low-lying coastal suburban parishes got swamped. Then the storm crashed into the Mississippi Gulf Coast. Then reports started surfacing that the levees and floodwalls had failed, and New Orleans was doomed.

When Damian Hebert got home from work that evening, the situation in New Orleans was getting worse by the hour, and his wife was still crying.

She told him about her day. He could see how it had affected her.

"You've got to do something," he told her. "Go volunteer. Do *something.*"

That was an odd suggestion. Margaret Hebert had never been the volunteering type.

"What can I do?" she asked.

"Just go show up at the Cajundome," her husband suggested.

The next morning, Hebert drove across Lafayette—university town, nerve center of South Louisiana's offshore oil and gas industry, "hub city" of the state's Cajun country—to the University of Louisiana at Lafayette's basketball arena. Evacuees from New Orleans had started arriving there prior to Katrina's landfall. She found some relief agency staffers and offered her services.

They turned her away. They didn't need her, they said.

Hebert was disappointed and more than a little conflicted. In no hurry to go back home, sit helplessly in front of the TV, and cry some more, she decided to hang around. Taking care to stay out of the way, she watched a real-life drama unfold before her eyes as displaced New Orleanians kept showing up.

As lunchtime approached, a man with a Red Cross identification badge called out, "I need twenty-five people to come and feed evacuees." The unsanctioned Hebert slipped in with the other volunteers and helped serve lunch. From then on, there was no turning back.

Conditions at the Cajundome grew more and more chaotic as hungry, tired, worried Katrina survivors continued to arrive by the busload. Eventually, more than seventeen thousand storm victims were delivered from New Orleans. The unfortunate brush-off that Hebert received that first day would prove to be an anomaly as the Cajundome staff worked diligently to coordinate relief efforts and embrace offers of assistance from throughout the Lafayette community.

Hebert found plenty to do at the shelter in her first few weeks there, either caring for people directly or assisting in the logistics of what was turning into a massive effort—building bathrooms, hosing down evacuees who showed up directly from the floodwaters of New Orleans, and serving meals. Many of the dazed storm victims arrived with nothing but the dirty clothes they were wearing and nightmarish memories of the ruined city they had left behind. Hebert made so many trips to the nearby Dollar General store to buy ladies' underwear and men's T-shirts for them—with her own money—that sympathetic store workers started busting the packages before ringing them up so they could sell them to her at a "damaged condition" discount.

"I probably did that for a month without even a name tag," Hebert would recall later. "I went ten hours a day, every day. This was the most rewarding thing I've ever done."

Or so she thought. Then Hurricane Rita swept across the Louisiana coast, and the little towns of lower Vermilion Parish got clobbered.

Consumed by the needs of the Katrina refugees now relocated to Lafayette, Hebert was oblivious to the extensive damage caused by Rita less than twenty miles from her own home. During the next week, a Red Cross worker told her how much the people were suffering down in Erath, where almost every house in the town had flooded. Stunned, but inspired, Hebert knew what she had to do. She filled her car with disaster clean-up kits—mops, brooms, and plastic buckets filled with scrub brushes, sponges, bleach, rubber gloves, and trash bags. Only then did she realize she had no idea where Erath was.

She called her husband.

"Look, they need help in Erath," she told him. "How do I get to Erath?"

Damian Hebert should have seen this coming. In the preceding four weeks, he had witnessed the beginning of a remarkable transformation in his wife, whose life in Lafayette to that point had been a comfortable one. She had worked for an interior decorator, worked for architects. She was a regular at the health club. She played a lot of tennis. And yet, here she was now, volunteering ten long, hard hours a day, wearing herself to a frazzle, doing whatever she could for storm victims who kept coming and kept needing. Clearly, Margaret was finding a new purpose in her life with each passing day.

"You've got to be joking," he responded.

"No, I'm serious," Margaret insisted. "I've got a car full of stolen things that I took from the Cajundome."

"What?"

"The manager knew that I took it. He just turned away because I'm going to Erath and bringing all these Red Cross buckets. They belong to Red Cross anyway."

Damian could only laugh and shake his head. "You know that road behind our house—Highway 339?" he said. "That will take you to Erath."

The drive took her about half an hour. She made her way to the town hall, which functioned as the command post for Erath's emergency response. She unloaded the donations then went back for more. She made the trip day after day, delivering clothes and buckets of cleaning supplies for needy Vermilion Parish residents who in many cases had lost everything in the flood. After several weeks, one of the relief agency workers she had met at the Cajundome saw her helping out in Erath and arranged for the Volunteers of America to hire her. Hebert was official now, with a nametag and a paycheck, but she would have soldiered on even without them.

What started as a trickle of relief supplies heading into Erath turned into truckloads of food, blankets, toiletries, diapers, and clothes, thanks largely to a massive email appeal sent out by the Rev. Wayne Duet, pastor of Our Lady of Lourdes Catholic Church there. Lisa Frederick, a director of religious education at the church and a volunteer in other community activities, stepped into the role of manager for the donation site the town set up at a city park. When donations threatened to exceed the storage capacity at the park, the owners of Champagne's Supermarket in Erath offered their ruined, vacant store, free of charge, as a distribution center.

Hebert was among the many people who spent long, uncomfortable days handing out relief supplies to local storm victims there. The gutted building was not insulated at that point, and there was no heating or air conditioning, so it was cold in the winter and stifling in warmer months. But the need was tremendous, and it wasn't unusual for the line of people seeking donations to wrap around the parking lot of the small-town supermarket.

"They were giving out commodities—mops, buckets, food," Hebert recalled. "We were out of paper towels, we were out of detergent. So that's it, we're out. Two days later, this eighteen wheeler came from we don't know where, filled with detergent, paper towels, cleaning supplies: simply a blessing. We had no idea, the stuff that we gave out. Brooms and mops, baby food and baby wipes, things that you need when you're working on your home and you're filthy and you don't have running water yet. We don't know where it all came from. It didn't come from Lafayette. It came from different states."

It wasn't just donations that poured into South Louisiana from far and wide to help the hurricane victims. The region was overrun with kind strangers who arrived from all over the United States and Canada to pitch in with the local recovery. Members of church and civic groups showed up for one to several weeks at a time, often spending their entire vacations there, helping one family or another do a little bit more to repair a damaged home or build a new one.

Mennonite Disaster Service established a base camp in Cameron soon after Rita and set to work rebuilding or repairing houses where most had been washed away or badly damaged. Three years later, when Hurricane Ike rolled around, Mennonite volunteers from across the United States and Canada were still rotating in and out, donating their time and skills. By the end of 2009, they had built fourteen houses in Cameron—dramatically elevated on high pillars to meet federal flood elevation guidelines and reinforced to the extreme to be able to withstand major hurricane-force winds—and made repairs on two dozen others.

Southern Baptist volunteers converged on many communities, repairing churches, serving meals, and helping the poor and elderly with home repairs. Many of the visitors—including a team with first-rate construction skills from Sevierville, Tennessee, and an entourage of forty-eight college and high school students on spring break from Fort Collins, Colorado—focused their efforts on Hackberry, where the First Baptist Church and the homes of many of its church members were wiped out by Rita and Ike.

The Catholic Diocese of Toledo, Ohio, buddied up with the Diocese of Houma-Thibodaux, setting up a long-term relationship that both parties hoped would extend beyond simple donations. Toledo church officials who visited bayou country following Hurricane Rita were struck not only by the storm damage they witnessed, but also by broader economic and social issues, intriguing cultural differences, and the common ground of faith. They determined to forge meaningful, lasting connections with the diocese ministering to more than 125,000 Cajun Catholics.

Indeed, when northern Ohio suffered its worst flooding in decades in August 2007, Catholics from the Houma-Thibodaux area were quick to respond. The Louisiana churches dedicated prayers in the first week of that September to their Ohio brethren. They collected monetary donations to help with the long-term needs of the northern flood victims. And, true to the practical nature of rural Cajun life, a nun from the Louisiana diocese's office of Catholic Social Services shared tried-and-true directions for cleaning mold from flooring and walls after floodwaters recede; the Toledo diocese promptly posted the "recipe" on its Web site.

Colleges across the country set up "alternative spring break" programs for students to venture down to Louisiana, instead of the Florida beaches, and spend their semester breaks assisting storm victims. Teams of AmeriCorps volunteers, eighteen- to twenty-four-year-olds who had committed to ten months of service in the National Civilian Community Corps program, pitched in on numerous hurricane recovery projects all across South Louisiana.

It would be hard to identify just how many different church groups, civic clubs, school organizations, and others ventured down to Louisiana to help out in the aftermath of Rita. In the Vermilion Parish community of Esther, though, Sherrill and Helen Sagrera felt like they saw them all. And they couldn't have been more grateful.

Hurricane Rita's storm surge wrecked their home, tearing all the bricks off the house and tossing their furniture and appliances around like salad fixings. The house was left standing—when they first checked it out after the water went down, a cow was standing in the front doorway—but it was a total loss, and the Sagreras had no flood insurance for the structure that had never flooded in the more than thirty years that they had lived there. Both sixty-five years old, they were reluctant to take on the challenge of building a new home, recognizing that they'd have to do it on the cheap and do most of the work themselves. But they couldn't bring themselves to buy a house "in town" and give up their land, the garden where they raised all their own produce, and the satisfaction of living so close to their daughter (next door), their son (across the street), and their church (around the corner). So they stayed put and started over.

They were in good company throughout lower Vermilion Parish.

"People mostly did for themselves," Sherrill said. "They didn't wait on FEMA. They didn't wait on anybody. As soon as that water went down, people started taking all that stuff out their house, throwing it alongside the road, trying to assess because according to the rules, if you didn't have more than 50 percent damage, you could fix your house as it is. If you were over 50 percent damage, you had to demolish the house. So they wanted to see exactly where they stood.

"The assistance itself from any kind of public entity or from anybody wasn't there. The people did it on their own. You could go down the road a week after the storm, most of the houses there had people working, cleaning up, rebuilding. It was amazing seeing people, as bad as the devastation was, putting things back together. It's tradition. They're used to doing for themselves. They figured, well, if we get help, we'll get help; if we don't, we're still going to survive."

To build a new home on their property, located seven miles from the open water of Vermilion Bay and eighteen miles north of the coast of the Gulf of Mexico, the Sagreras had to site it at least eleven feet above sea level.

"Before I could do anything, I had to go get an elevation survey," Sherrill said. "They come and put a nail in that tree and put that flag in and said, 'OK, you've got to go up at least this high.' I went up about four foot higher than I needed to."

Sherrill and Helen tore down most of the old house themselves, salvaging what little they could, and then set about scavenging building materials from six other homes in the area that also had flooded. End result: the roof of their new home came off one house, the wood-grain paneling came out of another slightly-flooded house, some of the flooring came out of a third house, and assorted fixtures came from others.

"I didn't have the money to go and buy, replace everything I had," Sherrill said. "I'm not too proud to use second-hand stuff, as long as it functions."

Sherrill got to know people running the United Methodist Committee on Relief program in Lafayette, and they arranged to send volunteers to help him and Helen build their house. They also got help from the Diocese of Lafayette, the Red Cross, and AmeriCorps.

"We had a different group every week," Helen said. "Each one of the teams paid their own way."

The members of one group bought four ceiling fans with their own money and installed them.

"It makes you feel good that people care," Helen said. "Seeing how many people from away from here were willing to come and help, that was something."

A bus rolled up one day, and when Sherrill saw a youth group emerge—ten girls, two boys—he thought to himself, "Aw, we're just going to be babysitting." His skepticism didn't last long. "Let me tell you something: they grabbed crowbars and hammers and you'd better get out of the way," he recalled. The young volunteers worked hard and well and kept up a cheerful attitude. When they needed a break, they went out into the street and played baseball with a two-by-four and a ball of mud. Sherrill and Helen could only look on in wonder.

Fourteen Ohio residents, almost all of them older than the Sagreras, spent a full week working on their house by day and sleeping at a campground in Lafayette at night. They built closets, installed insulation, hung drywall, set ceiling tiles, did some electrical work, and nailed paneling onto walls in all the rooms of the house. One of the volunteers, seventy-one-year-old Tim Holly, came from Xenia, Ohio, which had been devastated by a tornado in 1974. Remembering how people from all over the country contributed "everything from clothing to food to labor to prayers" after that natural disaster, he long had wanted to "return the favor," he said. He had made church mission trips to poverty-stricken areas of nearby Kentucky, but this was his first experience working with hurricane victims.

Group member Mary Kay Pyles, a retired school principal, lived part of the year in Cedarville and the rest of the year in Port Orange, Florida. In 2004, while she was in Ohio, her Florida home was hit by three of the four hurricanes to strike the Sunshine State that year. Devoted Florida neighbors took care of her home in her absence, cleaning the yard, patching the roof, hauling away debris and making necessary repairs to secure the home until she could return.

"When the hurricanes hit Louisiana in 2005, I felt that I needed to help as a way of paying back what I had received," said Pyles, then sixty-two. "The second reason is that I believe the value of our existence is in community and that each of us has skills and blessings to share. When the hurricanes hit Louisiana, I sent money, but didn't know how else to help. The answer came when members of the Cedarville Methodist Church were looking for ways to serve others in mission. Our pastor contacted friends of his in Lafayette, Louisiana, who made arrangements for us to come there."

Pyles marveled at the Sagreras' resourcefulness, their determination in the face of such personal setbacks ("how pleasant and gracious they were in the face of such great losses") and their trust in allowing so many strangers—many with no carpentry experience—to help build their home.

The elevated house is modest and haphazardly outfitted, with donated appliances and used cabinets salvaged from other houses and no kitchen pantry, but the Sagreras are happy there. As their first Christmas in the new home approached in December 2006, the couple received a box of Christmas presents from Ohio. The gift cherished most by the couple was a quilt signed by members of the volunteer group.

"We went down as fourteen individuals and came back as a team of friends," Pyles said.

That was a common result of the volunteer experience throughout storm-ravaged South Louisiana.

But few people across the post-Rita landscape of South Louisiana were more transformed than Margaret Hebert, who immersed herself in a community of battered but determined Cajun country folk much like those she knew as a child, before easing into the life of the big city. The little things that she found herself doing every day to help the poor people of Delcambre, Henry, and Erath get repairs made to their little old houses, or get into new, temporary housing, or just get a second-hand but useable appliance or set of clothes, gave her life new meaning—even if her friends back at the health club didn't understand.

Once she got her bearings in Erath, Hebert began to get more involved with storm victims on an individual level, not just as the masses of people filling a shelter or lining up for canned goods and diapers. When a clothing giveaway was organized at Dakota's bar, she took care to help adults and children alike sort through the piles of donated apparel to find items that each could use. She helped stage regular small-group gatherings called "coffee breaks" at fire stations, branch libraries, or other meeting spots in every community in lower Vermilion Parish, to update struggling residents on the latest developments in recovery efforts, answer their questions, and just listen to their concerns. She started visiting residents where they lived, or where they were working on homes they hoped to one day occupy again.

Many were elderly. Many were isolated, largely out of contact with the world beyond the small rural communities where they lived. Many had limited income, limited means of transportation and limited attention from the federal and state agencies that were supposed to be helping them.

"You get to know everybody when you're here this long," Hebert said. "One of my older clients I went to see, I asked how things were going. He had his leg crushed in an accident, now there's no money coming in, so two of his boys moved in. In order to make it, they've got three households in that house. He said, 'We'd have lost our house. We would be on the street had it not been for my two sons to come live with me.' These people here don't leave. They

Helen and Sherrill Sagrera, front row center and right, pose for a photo with some of the volunteers who helped them rebuild their home
in Esther.

Volunteers from Cedarville, Ohio, who helped Sherrill and Helen
Sagrera rebuild their house sent them this quilt as a Christmas gift.

live with their brothers or sisters or mom or dad. They stay. They're very close-knit. And there's a lot of raising grandchildren, helping that way, here that I don't see in Lafayette at all."

One family that she looked in on periodically—father, disabled mother, and children—lost their home in the hurricane and spent the next eighteen months living on a shrimp boat until someone donated a double-wide mobile home for them.

"These people, Rita people, for the most part did not wait for you to come by and give them something," Hebert said. "They did what they could to get it together."

Whether helping them get their utilities up and running, obtaining assistance in installing a wheelchair-accessible ramp, arranging donations of second-hand stoves and refrigerators for folks who had lost theirs in the flood and were living on canned goods, or helping to guide older people through the gauntlet of governmental programs that eventually reached the area, Hebert was affected by the work on a deeply personal level. The area had a significant concentration of older, French-speaking Cajun and Creole residents, and as she got to know them, the experience brought her back to the simple, happy time of her youth.

"I met some wonderful people from New Orleans, but nothing like these people here," she said. "I think because they're more like what I grew up with, and I didn't have that anymore. My parents were deceased, and it kind of felt like being back home here."

Hebert grew up in Mire, an Acadia Parish hamlet that doesn't even turn up on some state maps anymore. As a girl in an isolated rural community where little, if anything, ever happened, she spent lots of time behind the counter of Leger's General Merchandise and Plumbing Supply, the general store her father operated next door to Mire Elementary School on Louisiana 95.

"That was everything, that grocery store," she said. "I always had somebody to play with or visit with. We had to be nice to our patrons because that was our livelihood."

She sliced cold cuts and rang up sales, chatted up younger customers in English and the older ones in French and in the process, unknowingly, began to develop people skills that would serve her well throughout her life.

After arriving in Erath, one of the first people she visited in her new assignment was an elderly resident named Charles Thibodeaux. Meeting him for the first time at his home, she had a feeling that he spoke Cajun French, so she greeted him with, *"Bonjour. Comment ca va?"* (Hello. How are you?)

Thibodeaux replied, *"Ca va bien. Comment les caniques?"* (I'm well. How are the marbles?)

That caught Hebert entirely off-guard. She asked him, *"Mon caniques?"* (My marbles?)

He insisted, *"Mais, oui."* (Yes.)

Hebert still didn't follow. In English this time, she asked him, "Marbles? What do marbles have to do?"

A smile burst across the old gentleman's face, and he said, "Ah, you speak French!"

Taken aback, Hebert reminded him, "Well, I said *'Bonjour, comment ca va.'*"

Thibodeaux scoffed. "Anybody can say *that*," he said.

Hebert smiled too as she recalled the exchange months later.

"He was testing me right off the bat, because you have to know a lot of French if you know what 'marble' is," she said. "They're characters here. This has really been a blessing, meeting these people."

It wasn't hard for Hebert to recognize the ways in which meeting those people changed her outlook about a lot of things. Hamburgers, for instance.

Living in Lafayette all those years, she long had known that the best hamburgers in town could be found at Judice Inn, one of the city's cultural landmarks since 1947. In recent years, a competing body of thought began to tout NuNu's as the epitome of the local hamburger. That all went out the window once she paid a few lunchtime visits to Roland Viator at the Circle V Meat Processing Plant near Henry. But when she tried to explain that to her friends back in Lafayette, they wouldn't hear of it.

They remembered the Margaret who had been active at Red Lerille's Health & Fitness Club, known to everyone in Lafayette as Red's, and used to play tennis, lots of tennis. Now, not only was she working all day, every day, way out in the country, but she was eating at a slaughterhouse. Their response boiled down to, "OK, you've got to stop this."

To the contrary, though, the more she did for the hurricane victims of Vermilion Parish, the more she wanted to do—and the more disappointed she became with the ambivalence that seemed to pervade the big city that was so nearby. Like the rest of the country, even Lafayette seemed to her to be too preoccupied with the impact of Hurricane Katrina on New Orleans and its people to notice how seriously Rita had hurt its neighbors just twenty miles to the south in an adjacent parish.

She was grateful to see college students from across the country showing up to spend their spring break helping needy families with recovery projects, but it frustrated her that she couldn't gin up more such support from people closer to home. When nine high school boys came in from Lafayette to put in two days of work, she was appreciative but realized that they only showed up because they got into trouble and had to perform "service hours" as a consequence.

"When I have time to do my shopping and I see someone, I tell them what we're doing. You don't hear about Rita," Hebert lamented at one point. "Where are the things like furniture? What about one day of giving where everybody, in just one day, volunteers? What about getting a church or getting your family together and call me; I'll find you a family that you can work with, a family with children so you can see."

By the time Hurricane Ike hit, Hebert's work with the Volunteers of America had played out, and she moved on to a Lafayette-based affiliate of a national volunteer organization providing home repairs and modifications for low-income houses and communities throughout the region. The move was a logical progression for her, given the desire to help the needy that the hurricane experience of 2005 had brought out in her.

Hebert and her husband belonged to the Acadiana Riders Motorcycle Club. After her first several weeks on duty in and around Erath in Rita's wake, she helped the club members connect with needy families in the area who were staring down the prospect of a bleak Christmas. Many of the families had been forced out of their homes due to flood damage and had lost most of their possessions. With Santa in the lead, the club members spent one December day delivering presents to fifty-six children in the area.

"We had the best time, to see children that otherwise did not have Christmas," Hebert said.

And she never lost sight of what her volunteer experiences revealed to her: "There's a need here in our backyard."

A NEW SHERIFF
IN TOWN

Two weeks before Ike swamped Louisiana's coastal region, Terrebonne Parish tuned up with a visit from Hurricane Gustav. Even though it wasn't expected to be a particularly powerful storm, Gustav headed up the Gulf of Mexico directly toward Terrebonne Parish, so the local authorities called for a full-scale evacuation, and tens of thousands of parish residents retreated out of harm's way.

Like many down-the-bayou folks who make their living on the water, though, Sam and Patti Pellegrin rode out the storm on their boat. A partner in a family company servicing offshore oil rigs, Sam Pellegrin had moved his vessel well inland from their home on Bayou Little Caillou in Chauvin, and he and his wife were joined aboard by assorted family members who likewise had not heeded the directive to get out of town. As Gustav made landfall at Terrebonne Parish on the morning of September 1, they were safe, if stir-crazy, within the quarters of their solid, steel-hulled workboat.

After the storm blew through, Sam and Patti took a mid-afternoon drive from where the boat was docked in Houma to check their house for damage, fearing some of the pine trees from their front yard might have toppled onto the roof—or through it. A mere five hours after Gustav rumbled ashore with driving rain, funnel clouds, and sustained winds exceeding 75 mph, the breezes remained stiff but the sun was brilliant against a nearly cloudless sky. Making their way south along an otherwise isolated Louisiana 56, they got as far as Klondyke Road, about six miles from their home, where they reached a law enforcement roadblock. There, along with a couple of deputies, they found the new sheriff of Terrebonne Parish, Vernon Bourgeois.

Bourgeois had spent more than a quarter-century with the

sheriff's office, and anyone who knew him would not have been surprised to find him leading from the front in a time of crisis. In the three or so weeks between the run-up to Gustav and the immediate aftermath of soon-to-follow Hurricane Ike, Bourgeois was everywhere, conferring with other parish leaders here, meeting with the governor there, checking on parish jail trusties filling sandbags in Montegut, eyeballing the danger from downed power lines along the highway in Grand Caillou, and making his way into every storm-damaged community to gauge the impact and check on his deputies involved in rescue and recovery efforts. When it came to hurricanes, he knew that getting "into the field" and seeing any problem areas first-hand was better than relying on second- and third-hand information from others.

Spotting him at the roadblock that afternoon, Patti Pellegrin was encouraged. She and Bourgeois had known each other all their lives. They had grown up together in Houma's Barrios subdivision.

No matter. He wouldn't let them pass.

"C'mon, Verrrnon, let us go-o-o, pleeease," Patti implored him, smiling through her disappointment and drawing out each word in hopes of striking a sympathetic chord with her lifelong friend. "We *have* to see if we have a tree on our house."

"I can't," Bourgeois insisted, shaking his head and turning them away.

Patti tried to continue the conversation. Did he know their house, just past St. Joseph Catholic Church?

Yes, he knew it.

Could he tell them if any trees had fallen on it?

No, he wasn't sure.

Once more, with feeling: could he let them by, just long enough for them to see for themselves?

Not a chance.

The Pellegrins were frustrated at the time, but it later became apparent why the sheriff had been so adamant.

"They had power lines, trees, telephone poles all down on the highway, and you couldn't get around it," Sam said. "I understood after I saw that."

Vernon Bourgeois knew Terrebonne Parish, and its people, as well as anyone. He had spent years patrolling the bayou communities as a road deputy and even more years in the schools and churches and community centers running the Drug Abuse Resistance Education (DARE) program and as a liaison officer for the sheriff's office. Now, a mere two months since he had taken the oath as Terrebonne's new sheriff, the parish was about to endure a hurricane disaster that would take weeks to play out. The crisis would test Bourgeois' leadership skills, his endurance, and his character in ways he could never have imagined.

That he would find himself standing at a roadblock far out of town with his deputies on the day that the eye of a hurricane swept through the parish befit the arc of Bourgeois' twenty-six-year career at the Terrebonne Parish Sheriff's Office. There was a time, though, when it would have been more likely to find him trying to talk his way past a sheriff's roadblock rather than overseeing one.

Growing up, classmates couldn't help but look up to Vernon, because he was always one of the tallest boys in his class, but he also was friendly and funny by nature and had a knack for getting people to like him. As a member of the basketball team at Terrebonne High School, Bourgeois wasn't blessed with the natural talent of the starting five, but his determination earned him the role of super sub. He became the fan favorite, the reserve that everyone wanted to see take the court. The chant of "Ver-non! Ver-non! Ver-non!" would ring through the Tigers' gym at home games, and when Coach Ralph Bates would send in No. 35 to spell one of the starting forwards, the crowd went wild.

His ability to talk to anybody about anything led him to pursue a career in sales. Given the vagaries of the Terrebonne Parish economy, that was both a blessing and a curse. He went to work for Fleet Supply in Houma, selling batteries and filters and other automotive supplies to companies servicing the oil field throughout South Louisiana. When the oil industry was hit by a severe economic slump in the early 1980s, though, business slowed to a trickle, and Bourgeois' days as a salesman were numbered.

"I had just gotten married and the manager was trying to keep me on as long as he could," he said. "That ended up being a double-edged sword, because the people that got laid off in '80 and '81 got all the good jobs."

Bourgeois held on at Fleet Supply until 1982, but by the time he finally got pink-slipped, the unemployment rate in Terrebonne Parish was approaching 12 percent and there were no other jobs to be had, anywhere in the area. Production companies weren't drilling, so oil field work dried up for everyone, including the many companies that had thrived by providing and delivering supplies and services to the industry. The city of Houma, the rest of Terrebonne Parish, and most of the oil patch stalled amid the worst economic downturn since the local industry had first boomed decades earlier.

The situation left him "aggravated and depressed," he recalled. It got so bad that at one point, he walked into a paint store on Lafayette Street in downtown Houma and the guys behind the counter barked at him, "We're not hiring!" Insulted, Bourgeois took his business elsewhere. Looking back on the incident years later, though, he acknowledged, "I must have really looked desperate."

He looked desperate because he *was* desperate.

Seeing his parents one Sunday, he informed them he was going to make the rounds at Houma's Southland Mall the next day to apply for sales jobs at Sears and the mall's four shoe stores. His father, an insurance agent who served on the Terrebonne Parish Council, met him at the mall for lunch, and a chance encounter there would point his life in a new direction, one that he was initially reluctant to follow.

As they walked toward the A&G Cafeteria, Vernon Bourgeois Sr. spotted then-Sheriff Ronnie Duplantis, and he brought his son over and made an introduction. A pleasant conversation ensued, and Duplantis at one point asked the strapping young man what he did for a living. Bourgeois explained that he had lost his job at Fleet Supply and was there at the mall that day putting in applications.

"Well, come see me tomorrow," Duplantis told him. "You're a big guy. I'll hire you and you can start working the next day."

As much as Bourgeois needed a job, this wasn't at all what he had in mind.

"Sheriff, I don't want to be a cop," he told Duplantis. "I really don't. I've got friends who are police officers, and I'm not interested. I've heard the things they do and it never really excited me."

Undeterred, Duplantis set the hook and began to slowly reel in his catch.

"Just come and start," the sheriff replied. "If you find something else two months later, take it. I'll understand."

That was a Monday. On Tuesday, Bourgeois filled out an application at the sheriff's office. By Wednesday—this being a time

Surrounded by his family, Vernon Bourgeois takes the oath of office as sheriff of Terrebonne Parish, just weeks prior to the arrival of Hurricane Gustav and Hurricane Ike.

Terrebonne Parish Sheriff Vernon Bourgeois, at his desk in the Terrebonne Parish Sheriff's Office, Houma.

before extensive background checks became common practice—he was wearing a deputy's uniform and carrying a gun.

Despite his initial misgivings, Bourgeois came to enjoy law enforcement work—and he was good at it. Less than a year after being hired, he received a promotion. Before long, he was promoted again. When Jerry Larpenter became sheriff in 1988, Bourgeois was named a detective. He also spent several years visiting schools as a DARE officer, establishing a rapport with a cross-section of Terrebonne Parish youths and their parents that would serve him well years later when he sought elective office.

At the sheriff's office, tucked into the parish courthouse complex across from St. Francis de Sales Cathedral in downtown Houma, Bourgeois was effective in dealing with people, but he came to recognize that all the folks who turned up were seeking help with some problem, either their own or in many cases their children's. Out of the office, though—at the schools, at homeowners' meetings, amid the crowds along the parade routes for Mardi Gras, just on ordinary patrol—he thrived.

"When you're out seeing the community, people have lots of great ideas," he said. "I sponged a lot of good ideas off a lot of people. That's what I love about law enforcement—you learn something new every day. The challenges you have every day are awesome. Even if you're a deputy on the road and you're taking theft reports all day long, this theft report I guarantee will be totally different from the next theft report, and the one after that and the one after that. Some people don't have it. You see them last six months, a year, a year-and-a-half, and then they go on and do other things."

Then in 1996, a deputy named Chad Louviere committed one of the most horrific crimes anyone in Terrebonne Parish could remember. Early one October morning, the uniformed officer stopped a female motorist on Bull Run Road, pepper-sprayed her in the face, handcuffed her, and drove her in his patrol car to an isolated sugarcane field, where he raped her. A few hours later, he entered a bank on Grand Caillou Road and, armed with an assault rifle as well as his service revolver, he took his estranged wife and five other women hostage. In a rampage that lasted more than twenty-four hours, Louviere killed one woman, raped three others, and committed other crimes while holding the victims at gunpoint. He eventually surrendered to the SWAT team that had surrounded the bank. In time, he would be sentenced to death by lethal injection for his offenses.

Although the crime was solved, the incident torpedoed public confidence in the sheriff's office, poisoning the department's relations with law-abiding citizens throughout the parish. Residents were left to wonder about the integrity of other men and women in the same uniform that Louviere had worn. Many women were left to fear and distrust any officer that pulled them over for even minor traffic infractions or otherwise approached them. Morale within the sheriff's office fell as even the most dedicated deputies were viewed with suspicion by members of the public.

"It was a blow to this community," Bourgeois said.

But Larpenter knew that he had in Vernon Bourgeois a natural-born salesman. The sheriff set him to work both on the morale problem within the department and on the public relations problem out in the community.

Bourgeois started meeting with deputies, talking through the issues with them, reinvigorating them about their commitment to serving the parish and its people. He became even more visible in the community as he looked for ways to bolster the office's public image. He also took the lead in stepping up the sheriff's office's recruitment, because the Louviere case had called into serious question the procedures used to select and train deputies. An after-the-fact review of the gunman's work history revealed that Louviere had demonstrated psychological problems in his previous job with another law enforcement agency in the region, but his former employers had neglected to inform the Terrebonne Sheriff's Office of that when he applied for work there. Whether or not that absolved the Terrebonne Sheriff's Office of blame, it signaled that the department had to go beyond cursory reference checks when evaluating job applicants. With that in mind, Bourgeois began visiting high schools, community colleges, and universities to encourage desirable individuals to study criminal justice and pursue a law enforcement career with the sheriff's office.

"I started finding good people, and it really did make a difference, because we found people who really did want to do this for a living," he said. "You want to make it a career—you don't want to just make it a job."

Having proven himself time and again as a go-to guy, he eventually worked his way into the rank of major and the role of the sheriff's top assistant. As Larpenter began to think about retirement, he started grooming Bourgeois to replace him. Among the many responsibilities Bourgeois now embraced, none was more critical than running the agency's operations during hurricanes and other natural disasters—planning for a range of contingencies and then mobilizing the 330 employees to most effectively keep the peace and protect lives and property during emergency situations.

"It's a lot of work. It'll drain a person. It'll age a person quick," Bourgeois said. "You're always responsible on a day-to-day basis for more than one hundred thousand people, but in a hurricane situation you're involved in lives and property and many things being taken care of at that time."

When Terrebonne was hit by the storms of 2005—Katrina's slap on August 29, then Rita's body-slam on September 24—Bourgeois was put in charge of coordinating the sheriff's office's response.

"Rita came in, we kept getting high tides, we kept getting southerly winds, and the water came and stayed for a long time, which caused more damage," he said. "If water comes in and out of a place, it's not that bad. I mean, sure the stuff's wet, but that's it. When it comes and stays and sits, though, it causes major damage. That's what happened across the parish."

The area had not taken a major hit from a hurricane in many years, so many thousands of residents did not evacuate, either failing to recognize or choosing to ignore both the immediate danger and the lingering problems that a storm's impact could cause.

"I think from years past, they just always stayed," Bourgeois said. "You've got your ones who just anchor themselves in their homes: 'I'm safe. My home stayed here for four hurricanes. It'll survive this one, too.' They had that mentality, and they probably thought Katrina was a one-time thing and that was it."

Consequently, when Rita swept through and left havoc and devastation in its wake, the challenge to law enforcement agencies was compounded by the presence of too many people who should have evacuated from the flood zones but didn't, or who tried to get back into the flood zones before it was safe to do so.

"A lot of people didn't evacuate during that time," Bourgeois said. "There were people running around, there are things that people don't know that go on behind the scenes. The power line

"Vern and Mr. Bobby": Sheriff Vernon Bourgeois greets Gov. Bobby Jindal on one of his many visits to Terrebonne Parish during the Gustav/Ike crisis.

people—they do triage first. In hurricanes past, it takes two or three days to do triage. If you don't plan what structures need to be brought up first, you're back-stepping many times. You've got to find areas that have the most damage or those with the most critical infrastructure, repair those first and then you go on from there. For many hurricanes, it took two and three days of triage, just for the electric company, because you have people in your way as you're going from place to place."

The post-Rita year of 2006 was an unsettling one in Terrebonne Parish. Rita had flooded an estimated ten thousand homes from east Houma through the bayou communities stretching toward the Gulf. As public, individual, and volunteer recovery efforts began in earnest, residents and their elected state and local representatives talked in increasingly dire tones about the need for substantial flood protection measures. Ideas for keeping out the next storm surge were plentiful; ideas for how to pay for it, not so much. All the while, everyone kept a nervous watch on the weather forecasts, fearful that the arrival of another hurricane before the parish could dry out, fix up, and start over from the last one would mean unprecedented catastrophe, from which the area might never recover.

But 2006 proved to be hurricane-free for Louisiana. So did 2007. The only squalls that year were of a political nature, as voters in the fall elected a new governor and a variety of other state and local officials. As expected, Larpenter decided not to seek re-election then, choosing to retire after twenty years as sheriff. Bourgeois, by now holding the rank of major as the director of operations for the sheriff's office, was one of four candidates to enter the race to succeed him. The election was held on October 20, and the sheriff's understudy outpolled all other candidates combined to win without a runoff.

"It felt good that the people of the parish had enough confidence in me to elect me," Bourgeois would say later. "I'm sure my being here that long helped—I had been here twenty-five years at the time."

Most of those elected on the state and local levels took office in January 2008, but sheriffs' four-year terms in Louisiana start on July 1 following the previous fall's elections. That gave Bourgeois eight months to plan the switch from Larpenter's administration to his own. The transition was a smooth one, but within weeks of Bourgeois' taking the oath of office, a hurricane appeared on the far horizon, and things got interesting in a hurry.

As soon as Gustav formed as a tropical storm on August 25, far out in the Caribbean, the new sheriff discovered that everyone expected him to be an instant authority on the matter of hurricanes. Overnight, his honeymoon period as the newest parishwide elected official came to an abrupt halt, and encounters with everyone in restaurants, in the halls of the courthouse or anywhere else went from "Congratulations, sheriff!" and "Hiyadoin', sheriff?" to "What's that storm gonna do, sheriff?"

Five days out, on August 27, Hurricane Gustav had just pummeled Haiti and was still on the far side of Cuba, but the National Hurricane Center forecasters' best guess had it crossing the Gulf of Mexico and coming ashore at the southeastern edge of Terrebonne Parish. The new sheriff took that as a good sign—hurricane forecasters were never *that* accurate—and he told everyone so.

"Trust me, it's not coming here," he told people with as much good-natured assurance as he could muster, while remaining cognizant of any such hurricane's potential for destruction if it *did* approach. "If they've got it going to Terrebonne Bay, trust me, it's not going anywhere near Terrebonne Bay."

The next day, Gustav was nearing Jamaica. Oil companies were wrapping up their evacuation of offshore rigs. The Weather Channel experts continued to center their "cone of probability" on Terrebonne Parish.

Bourgeois remained unconvinced. "Trust me, people. Don't worry about it," he told acquaintances that he met on the street. But he didn't go to the media and say that for the record.

Three days away, the hurricane churned through Jamaica and headed for Cuba. Forecasters continued to plot a course up through the Gulf of Mexico to the coastal area south of Houma. The sheriff relented and started telling people, "OK, everybody start getting ready, because it's actually doing what they called. The weather guys are finally getting one right, maybe. I still hope they're wrong, but maybe they'll be right."

Two days before U.S. landfall, Gustav struck Cuba as a Category 4 hurricane, with the most violent wind gust measured at 211 mph. By now it was obvious to all in Terrebonne Parish that the meteorologists had been correct all along and the storm was coming right at them. A mandatory evacuation order was issued, and parish residents started moving out to safer locations. Many loaded up their vehicles and took off for North Louisiana and neighboring states. Hundreds filled shelters that opened at schools or other public buildings in and around Houma. Many others boarded buses ordered up by authorities, and they left for points undetermined. Inmates at the parish jail, situated in flood-prone Ashland along Bayou Grand Caillou, were transferred to the state penitentiary at Angola.

Sunday, August 31, the day before the storm hit, authorities made their final preparations while many residents who hadn't rushed to evacuate earlier faced the last-minute decision about whether to stay or go. Weather conditions across the Gulf beat down some of the storm's intensity as it approached Louisiana.

The eye of Gustav came ashore near Cocodrie on Labor Day morning, a Category 2 hurricane with sustained winds of 105 mph. For all its bluster, it proved not to be a very wet storm. It brought little high water, except in the southernmost areas closest to the coast, and didn't dump much rain through the parish. The winds, however, tore up homes, businesses, and several schools. It toppled thousands of trees and hundreds of utility poles. Roads were left inaccessible, not from floodwaters, but due to trees, power lines, and other debris strewn across them. Power was out everywhere. Communications were severely hampered.

It was a direct hit, but Bourgeois didn't panic. As a sheriff's office veteran, he had worked many storms before—Juan in 1985, Andrew in 1992, Georges in 1999, Isidore and Lili in 2002, and Katrina and Rita in 2005. As his responsibility steadily increased, each storm had taught him something. Now, he was in charge, and the leadership team he had assembled at the sheriff's office was poised to meet Gustav head-on.

"I had worked through so many hurricanes, did so many various things before, during and after hurricanes, that I'm confident when a hurricane comes," Bourgeois said. "It worries me that people could lose their lives—I don't like that. We do the best we can to tell people to evacuate, to express the dangers that are coming toward them. But operations before, during, and after a hurricane don't bother me. And it's not just me. We're comfortable here with about twenty people who have been here for a long time, as long as me. They know what do to first, second, third, fourth, fifth, in the right order. I noticed, when I thought something needed to get done, I would call the person in charge of that section, because I designate authorities—and they'll say, 'Sheriff, I just started that twenty minutes ago. We're half done now.' Everybody I called, that's what I got. Or, 'We're just about to finish this, and that's next on the list.' I got that a lot."

At the parish government's emergency operations center, though, things weren't going so well.

Attorney Michel Claudet had taken office as parish president in January, having been elected as a reform candidate, a political novice promising "a new way" for the future of Terrebonne Parish. After eight months of smooth sailing as parish leader, Claudet faced his first serious challenge when Gustav blew through, and he was unprepared for the crisis.

Wind damage from the hurricane toppled about 75 percent of the power poles across the parish, along with many radio towers. There was no electricity. Cell phone service was wiped out; most land-line telephones weren't working either. The loss of electric-

Sheriff Vernon Bourgoeis in the Sheriff's Office emergency operations center. A satellite image of Hurricane Ike is projected on the monitor behind him.

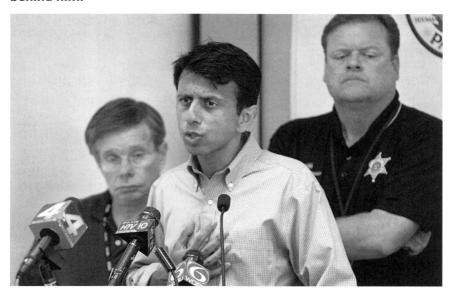

Gov. Bobby Jindal, flanked by Terrebonne Parish Pres. Michel Claudet, left, and Sheriff Vernon Bourgeois, discusses Hurricane Ike during a tour of Terrebonne and Lafourche parishes.

ity crippled other public services—the hospital shut down, there was no running water, no flushing toilets, and no working gasoline pumps. Meanwhile, parish personnel had minimal communication with each other, much less the outside world. Evacuees anxious to learn about conditions back home couldn't get phone calls through to parish offices, couldn't find current information on the parish government Web site and couldn't get meaningful updates from the media because parish officials weren't sharing information with the media either.

Claudet had appeared to be micro-managing the parish's storm preparations before Gustav made landfall with good intentions but without a clear understanding of what needed to be done, when it needed to be done, and how it needed to be done, especially in light of different possible scenarios such as extreme flooding or, heaven forbid, a total loss of communications. Now, the novice chief executive was overwhelmed. Confusion ruled the day.

Gov. Bobby Jindal came to town the day after Gustav struck to meet with the parish president, Terrebonne Parish legislators, and other parish officials, including the sheriff, to get an update on the local situation and determine what help the state could provide. A fast-talking, take-charge whirling dervish whose air of confidence and competent authority would serve hurricane-battered Louisiana well that September, Jindal was surprised to find the parish government still unable to gain any traction toward recovering from the storm. By the time the meeting ended, a weary Claudet had given up control of the parish's hurricane response—and turned it over to Sheriff Vernon Bourgeois.

Claudet would be criticized by many, first for failing to mount an effective storm response plan, and then for abdicating his authority in the heat of battle to the sheriff. Frustrated residents even launched a recall drive, although they failed to obtain enough petition signatures to force an election. For his part, though, Bourgeois stayed out of the political fray, focusing on the tasks at hand and giving Claudet the benefit of the doubt.

"The parish president had a little problem," Bourgeois said with considerable tact. "He's a good guy. He's just not great in emergencies.

"Honestly, I think he saw us get things in motion, and he didn't know what to do first, second, third, or fourth, which is understandable because he's never done it before. I've done it several times—more times than I like, don't get me wrong. I don't like it, but I know what comes first, second, and third. I know that if you don't do this first, it's going to come back and hit you in the head three or four or six hours later. If somebody else got elected parish president, they wouldn't know what to do next. That's just the way it is. And that gives people fear. When somebody wants you to do something you haven't done before, it makes you feel uncomfortable, and I'm sure that's what his feeling was. He saw that I was flowing well, I was getting things done, and he simply asked if I would take it from this point on and carry the ball, and get it finished. That's what I did."

Most residents had complied with the mandatory evacuation order. With so few people on the streets, deputies, Houma police, state troopers, and National Guard troops found it easy to keep the peace, assist utility workers, and do whatever else needed doing.

"It was like a ghost town. It was awesome," Bourgeois said. "All we had was us and the linemen, going from area to area to triage. We were protecting people's property, there was livestock that we had to move around, basically that was it. Our deputies could patrol, they could see if somebody was there that didn't belong there. It was real easy to identify them. I was really impressed with the public, and I announced that with the media immediately, praising them for what they had done, praising them for their trust in us to protect their property while they're gone. That's what people worry about—their house being burned down or vandalized, things like that, their car, house, property, whatever."

It helped, too, that the National Guard units sent to assist in Terrebonne Parish were from Louisiana instead of some distant state as had occurred in years past. The soldiers and commanders understood the culture and the local way of doing things, and that made communication with the local authorities easier and simpler. In some cases, the soldiers had been there before, so that made logistics easier, too. They knew where the mall was, where Coteau Road was, where Cocodrie was. In years past, a deputy might have to ride along to show the out-of-state guardsmen the way, or they'd be given a map and detailed directions and sent off, only to arrive two hours later when the trip should have taken them only fifteen or twenty minutes.

Unlike the parish government, the sheriff's office had not lost its in-house communications system, so deputies were able to contact each other without interruption. When the phone systems came back on line, though, staffers at both the parish government and the sheriff's office were crushed with phone calls from evacuees desperate to know what had happened back home and when they would be allowed to return. Bourgeois and other officials urged them to stay away, warning them that they were better off staying wherever they were than returning home to no electricity, no water, no sewer service, no medical care, and no open stores or gas stations. Some came back anyway, but many heeded the advice. That was fortuitous, because Hurricane Ike turned up the very next week, bringing it with it some of the worst flooding Terrebonne Parish had ever seen.

Ike inundated lower Terrebonne Parish, flooding an estimated thirteen thousand homes and other buildings and cutting off all the bayou communities from the parish's hub at Houma. While Gustav had caused tremendous distress due to the loss of services, the widespread flooding spawned by Ike presented serious threats to life and property. Dozens of residents who ignored the evacuation orders had to be rescued from their flooding homes by boat as Ike's storm surge swamped low-lying communities. Two people died in the flooding.

"We were hoping not to have a repeat of Rita, but we have that and worse," Bourgeois announced at a news conference in Houma the morning after Ike struck.

Deputies sustained their search-and-rescue operations while the water was up, then shifted to other familiar storm-aftermath duties once the flooding receded. It was worse this time, but they knew what to do. They'd done it before.

Governor Jindal visited Terrebonne Parish seven times within a span of nine days. Bourgeois could not have been more grateful for the governor's assistance, his concern, just his presence.

When Bourgeois met him on the first occasion, Jindal shook hands with him and said, "Hey, sheriff, how are you doing?"

Bourgeois responded, "Governor, just call me Vernon. I'm comfortable with that." And he was. Except at crime scenes or important functions where protocol dictated that he be addressed by his title, the informal Bourgeois preferred that everyone call him Vernon.

The "Hey, sheriff" / "Call me Vernon" exchange replayed at the governor's second visit, and his third.

The fourth time Jindal showed up, Bourgeois greeted the governor with, "Mr. Bobby, good to see you, man." The governor smiled and said, "Vern, how are you doing?"

No matter that no one ever called him "Vern," or that Jindal, at thirty-seven, was twelve years his junior; from then on, it was Vern and Mr. Bobby.

As the parish began to dry out, fix up, and move on from the one-two punch of Gustav and Ike, someone spray-painted a modest sign on a weathered sheet of plywood and mounted in on the side of the road near Evergreen Junior High School. It said,

> Thank God
> we elected
> Vernon B
>   &
> Bobby Jindal
> God bless

Bourgeois framed a small snapshot of the sign and mounted it on a shelf behind his desk. In an office otherwise decorated with an impressive array of New Orleans Saints and LSU football mementoes, the photo was one of his most cherished possessions.

"It says a lot, doesn't it?" he said. "And the person probably couldn't spell my last name—they just put 'Vernon B.' That's all right with me."

HOME IS WHAT YOU MAKE IT

It was a typical morning along the Mermentau River when Charlie Theriot and his brother Frank set out to do some hunting.

There seemed to be no one else around for miles. The air was clean and fresh. The area was teeming with wildlife but a quiet serenity pervaded the landscape, punctuated only by the gentle rustling of leaves or the trill of the occasional red-winged blackbird.

Charlie was closer to his younger brother Frank than any of his other siblings. Just two years apart in a family of four boys and three girls, they had grown up as playmates. As they got older, they became outdoor buddies, fishing and hunting together. When their father, a farmer and trapper, died at the age of forty-one, Charlie assumed much of the responsibility for providing for their mother and the rest of the family. He was fifteen.

Two years later, that circumstance had brought Charlie and Frank even closer. Frank was 15 by then, able to help Charlie with the farming. Hunting was no longer just a pastime for them; it was a necessary means of putting food on the table.

On this day, the teenagers were making their way through the familiar area north of the river when they reached a fence line. Preparing to cross the fence, Frank put his gun down haphazardly, not realizing it was loaded. The weapon discharged, shooting him in the stomach.

Alone there, miles away from any help, there was little Charlie could do. As he held his younger brother, talking to him, trying to stop the bleeding and comfort him, Frank died in his arms.

The year was 1929. The country was on the verge of the Great Depression. But for Charlie Theriot, as devastating as it was to lose his little brother, his best friend, right before his eyes, depression was not an option. There were still crops to raise, traps to run. His mother and his sisters still at home were depending on him. The experience steeled him to carry on, to muster all the stubbornness it would take to survive.

Well into the twenieth century, survival remained a challenge in the rugged isolation of the chenier plain. Such extreme reaches of lower Cameron Parish still were accessible from the outside world only by boat; if outsiders knew about the place at all, chances are they dismissed it as uninhabitable. The Theriots grew cotton, even though it was hardly worth anything back then—only five cents per sack. They grew sugarcane and made their own cane syrup. They planted corn and sweet potatoes, and they trapped mink and muskrat. Down there, back then, life was hard, and anyone that endured it came out tough.

As time went on, Charlie came to epitomize the standard-issue Cameron Parish way of life: live off the land, dodge the hurricanes, and pass the torch.

On April 21, 1932, at the age of twenty, Charlie married Macilda Miller, then sixteen. The couple settled in Grand Chenier, and soon they were rearing three children of their own in a sturdy old farmhouse built with wooden pegs by Macilda's father and brother in 1912, the year Charlie had been born. To support his family, Charlie did what he knew—farming and trapping, hunting and fishing—and over time he took to raising cattle as well.

Life on the chenier was lived in harmony with nature, as it had been for generations, but nature could prove discordant during hurricane season. When Hurricane Audrey took aim at Cameron Parish in 1957, Macilda's parents, other relatives, and lots of grateful neighbors joined the Theriots in their sturdy farmhouse—three miles from the Gulf of Mexico shoreline—to ride out the storm.

"We had twenty-six people in the house for Audrey," Theriot recalled fifty years later. "People from five and six miles away came to that house."

When the ground floor began to flood, everyone inside moved upstairs.

"There was a family across the road from us, and he brought his wife and son to the house," said daughter Lidian Richard, who was eleven at the time. "But he went back to his house to open the cow-pen gate. He said he didn't want his cattle to be trapped if the water came up—and it came up so fast that he drowned. He never made it back."

"I remember after the storm," Richard added, "some of the men left our house and went and rescued some people that were up in the trees, holding on to branches."

Survivors came to respect all the storms that would follow in ensuing years.

The old wooden house withstood the violent crashing of the waves as the hurricane made landfall, but the incessant pounding of the breakers atop the storm surge compromised the structural integrity of the second story, and the house had to be torn down afterward. The Theriots built a new house—five bedrooms, one story—on the same location.

The hurricane of 1918 had blown the roof off Charlie's boyhood home, but Audrey was the one everybody remembered. By the time he reached his nineties, Theriot figured one galvanizing event like that was enough for him, his family, and his community in his lifetime. Hurricane Rita had other ideas.

The Theriots evacuated for Rita like everyone else, and they spent about two and a half months with their grandson, Charles Hebert, and his family in Youngsville, near Lafayette, before they could return to Cameron Parish.

Back in Grand Chenier, there was nothing left of their house but debris scattered across the narrow strip of land and into the marsh. Charlie and Macilda had seen this before, right after Hurricane Audrey.

"What was left was so messed up with mud and trash, they had to dig a hole and a bulldozer just pushed it all in," Charlie said. "Couldn't save nothing."

But it was still home.

After a family friend made temporary arrangements for them up in Grand Lake, it didn't take long for Charlie, now ninety-five, and Macilda, ninety-one, to decide on a long-term plan.

"I had $34,000 worth of dirt brought in to raise up the property," he explained in a manner that suggested the decision was so obvious it needed no explanation. "Then we put a concrete slab, then we bought a double-wide trailer, and we're going back."

Charlie couldn't envision living anywhere else by choice. The land was rich, and even though full-scale farming wasn't prevalent anymore, vegetable gardens were plentiful. The surrounding waterways abounded in shrimp, crabs, and fish. And there was something more about the place, something that they took for granted until 1957.

After Hurricane Audrey had wrecked Grand Chenier, the Theriots had moved in for a time with a nephew in Lake Charles. Charlie couldn't abide that city living, where folks didn't grow their own vegetables and didn't even know their neighbors.

"People in town don't know who lives next to them," he scoffed. "In the country, you know people for miles and miles. You know them all."

As for Grand Chenier, he said, "You can't find a better place."

In late June 2007, just days before the fiftieth anniversary of Hurricane Audrey's landfall in Cameron Parish, Charlie and Macilda sat in the living room of their temporary quarters in Grand Lake and discussed their impending move back to Grand Chenier. They had hired Roy Bailey Construction to put in their house pad, a mound of dirt high enough to keep their home out of harm's way when the next big one hit. The double-wide trailer they had acquired was hooked up and outfitted. A driveway had been installed on the house pad so Charlie, who had just gotten his driver's license renewed and was good-to-go for another four years until his ninety-ninth birthday, could drive up to the front door. All that remained was for the furniture store in Lake Charles to deliver mattresses for their new bed, and then they would be moving in, in a matter of days.

"I thought I'd never live in a trailer house, but it's everything we need," Charlie offered. "It ain't fancy or nothing, but . . . but . . ."

Macilda finished the sentence for him: "A home is what you make it."

Everyone in their extended family—the two surviving children, the six surviving grandchildren, the nine great-grandchildren, Charlie's three sisters, and the beaucoup nieces and nephews and godchildren—realized how important it was to Paw-Paw Charlie and Maw-Maw 'Cilda that they get back to what they called "the old home place." But no one understood better than Lidian's grandson, Ryan Bourriaque.

A student at Louisiana State University at the time Rita struck, Ryan was an anachronism among his peers. He was an old man disguised as a twenty-two-year-old, a throwback to old-time ways and old-time values, proud to be a sixth-generation descendant of a family that had lived fulfilling lives on the same narrow strip of land in Louisiana's isolated southwestern corner for more than 150 years. And although he set off for college with every intention of making a life for himself elsewhere, and earned two degrees in the process, the tug of hearth and home would set him on an unexpected, and unexpectedly fulfilling, career path back to Cameron Parish.

Growing up in Creole, Ryan was heavily influenced by the older members of his family, especially his grandparents and great-grandparents. When he was in sixth grade, his paternal grandfather died of a heart attack. His grandmother, Viola Bourriaque, had been suffering with cancer for several years, and now she had lost her husband; soon, Ryan's parents sent him down the road to live with her. He proved to be a big help around the house, but the companionship that he shared with his grandmother over the next four years would dramatically impact him in ways that he wouldn't fully appreciate until years later.

"I moved in with her, and that's when I learned how to cook, how to clean, how to wash dishes, how to wash and iron my own clothes, and still take care of the yard and whatever she needed taking care of," Ryan said.

Maw-Maw G, Ryan always called her. When he was a little boy, his family wanted him to call her Maw Maw V—for Viola—but he had trouble making the "V" sound. It came out "G" instead, and the name stuck. Devout in her Catholic faith, quiet in her suffering, Viola grew very close to her grandson-helper.

"We took pilgrimages overseas three summers in a row, nineteen days at a time," he said. "We went to Fatima, Lourdes, Frankfurt, Venice, Florence, Rome—just wonderful trips. Me and forty other people from South Louisiana, and the person closest to me in age was forty-seven, which didn't bother me because I've been around grownups my whole life. Moving in with her, I became attached to her friends, people of her generation. They became my guardians, so to speak. They had vested interests in what I did with my life, who I was associating myself with, what did I want to become. To this day, it's the exact same."

As Ryan continued to live with his ailing grandmother, he found out who his true friends were.

"Here I am a junior in high school, and I'm not going out drinking and doing all this other stuff," he recalled. "I had a core group of friends that would come and meet me and sit there and play cards with us instead of going and doing what young kids do. And they would never blink an eye. They called her Maw Maw G just like I called her."

Viola succumbed to cancer in 1999, while Ryan was in eleventh grade. He moved back home and, influenced by the years spent with Viola, he drew closer to his maternal grandparents, Lidian and her husband Lester Richard, and to the patriarch and matriarch of his mother's family, Paw-Paw Charlie and Maw-Maw 'Cilda.

To the extent that Ryan indulged in under-age alcohol consumption, much of it occurred at the Theriots' kitchen table.

"He and I would always have a drink together and we'd visit," Ryan said of his great-grandfather. "That's what I like to do. I like to learn about things like how much he paid for his first truck. He paid $450 for his first vehicle and had to work three months to pay off his debt. He only had $150 to put down on it."

A neighbor loaned him the other $300. Charlie worked the man's fields for three months to pay him back.

"It was stories like that," Ryan said. "We'd fix a drink and talk. Every time I'd visit with him, he was ready to talk, he was ready to have a drink. I hope deep down that he realized my friends and I didn't always go there just to drink his whiskey. We went there for the words that were spoken over the drinks. I think he knew that, but I wish I had made it more clear to him that I was the lucky one. He would drink my friends under the table and would still be walking around ready to have another one and they were like, 'Can we go home? Can we stop?' "

Even at his young age, Ryan could sense that his great-grandfather was the last of his breed. He didn't begrudge the old man his stubbornness; rather, he admired him for it.

"He just had to live through things that you will never comprehend. And you don't want to comprehend them. That's the whole thing. You take it for granted," Ryan said. "I like that about him—I like that he's set in his ways, that he's particular, that he's going to tell you what's on his mind.

"Maw-Maw 'Cilda was the equalizer. She was the peacekeeper. But him? If you crossed him, and he thought you did it intentionally, he was going to call you out on it. He didn't need to mince words. He knew what he wanted to say and he said it. And I appreciated that. I loved that. Fortunately, and unfortunately, I have some of him in me. Every one of us had a bout with his stubbornness, the way he was so particular, and we've all had it out with him at one time or another, everyone in the family. But in the end, he was still a loving father, a loving husband, a loving brother, a loving uncle, grandfather and great-grandfather, and in the end, he was our friend."

If anyone in the family ever doubted Charlie's devotion to his family, that was set straight around the dinner table at Christmas every year.

Back in 1967, Lidian and Lester had a baby girl. They named her Jennifer. At the age of three months, the child died of sudden infant death syndrome, a week before Christmas.

Every year thereafter, when the family gathered for Christmas dinner, everyone would say grace, and Paw-Paw Charlie would put his head down, think about that infant granddaughter and start to cry.

"I just remember that little girl sitting there while we were decorating the Christmas tree, and the next day she passed on," he would say. She was like a little angel, he said. Decades later, the image remained branded in his memory.

Ryan graduated from high school in 2001 and enrolled at LSU, still devoted to his hometown and his extended family but eager to chart his own course in the world. He buddied up with about a dozen other students from Cameron, Creole, and Grand Chenier who had graduated South Cameron High within a two-year span, and that group expanded to a network of thirty to forty friends who hung out together, listening to music, cooking, visiting, just having a good time. After four years of studies, tailgating, and campus life at the state's flagship university, Ryan earned his bachelor's degree in 2005 and went right to work on a master's in the field of resource economics and policy, still not knowing where life would take him.

One Tuesday in mid-September that year, Ryan was sitting in the hall of a classroom building on campus, waiting for the start of his Intermediate Economics class. There was a test scheduled for that day's class, but rather than getting in some last-minute studying, Ryan was clearing his head by reading the latest issue of the Cameron Parish *Pilot*.

A classmate, Megan LaGrange, was sitting across from him. She asked him what he was reading.

"This is our local newspaper. It's the Cameron *Pilot*," he told her. "We get it once a week and basically the first thing everybody looks at is the obituaries and the sheriff's report, just to make sure none of your family members are in there."

Megan laughed and said, "That's funny, we have a paper like that too, in Patterson," her small hometown in St. Mary Parish.

In class that day, the professor assigned some homework problems designed for students to pair up and work together. Ryan partnered with Megan, who had a campus job at the School for the Coast and the Environment, near where Ryan worked at the university's Sea Grant office.

That Thursday, as Hurricane Rita was churning through the Gulf of Mexico, Ryan skipped class to drive home to Cameron Parish and help his family prepare to evacuate. He and his mother boxed up all of Charlie and Macilda's personal items, including framed photographs from around the house and the journals Charlie had been keeping since the 1950s, loaded up what they wanted to take with them, and stashed everything else in the attic of their house in Grand Chenier. Back in Creole, Ryan packed for himself. That didn't take long.

"I had yearbooks, I had a family Bible that my Grandma Viola had willed me, I had two statues that she had given me, and photo albums. That was it," he said. "My mom said, 'That's all you're taking?' I said, 'You know, it's kind of philosophical, but it's real to think that you know you could be losing everything. Of everything in this house, all I really want, all I really need, is what can fit in this copy paper box.' "

Megan called him Friday to let him know she had copied for him the notes she had taken from Thursday's class, and they talked about getting together to go through everything after the storm passed—not anticipating that Rita would cause such upheaval. Ryan evacuated with his family to a camp at False River, northwest of Baton Rouge, and it was the middle of the next week before he caught up with Megan back in Baton Rouge.

"She was crying, thinking about what had happened to us," he said. "She had everything organized for me. She had all the notes typed up for me, the homework assignments that she had worked

Charlie and Macilda Theriot.

Ryan Bourriaque sits in the open field of the family homestead in Grand Chenier.

on as much as she could."

They started spending more time together, got to know each other better, and a romance bloomed.

"Out of all the things that Rita took away from us," Ryan said, "Megan is one of the things that I got back."

Once he earned his graduate degree in the spring of 2008, Ryan faced an array of options. He was heavily recruited by a Catholic university in Minnesota to enroll in law school there. LSU offered him a graduate assistanceship to continue his graduate studies toward a doctorate. Based on his academic background and his student work with the state Department of Natural Resources and LSU's Sea Grant office, a private consulting firm offered him a position.

He applied to the law school and pondered that opportunity for a few months, but eventually decided he didn't want to leave Louisiana at a time when he had just become godfather to his cousin's baby, as his younger sister—his only sibling—was about to enter her senior year in high school, and while the elderly great-grandfather who meant so much to him was still around. Left with several closer-to-home options, he was still struggling with the decision a month after graduation. Then he got a visit from someone with the Cameron Parish Police Jury. The parish government back home was preparing to create an assistant planner's position in its planning and development office, to oversee hurricane recovery projects.

He interviewed for the position, returned to Baton Rouge, and was offered the job two weeks later.

"After laboring for months on end, on trying to decide what I wanted to do, the decision was made for me," he said. "Everything just worked out perfectly. I didn't have much planning experience; I had taken land use and policy classes, but as far as planning a community, I hadn't had that experience. But I kind of found a niche once I started working."

He could have made more money in the private sector, or a few years hence with a law degree, but that didn't matter.

"For me, it was a good salary, because I don't need a lot," he said. "The happy people are the ones who don't want a lot, that's pretty much what I've learned."

Ryan returned home and immersed himself in the business of helping to guide Cameron Parish's recovery from Hurricane Rita. He had hardly gotten his bearings when Hurricane Ike struck,

Actor George Clooney greets local residents following a ceremony to mark the rebuilding of South Cameron Memorial Hospital, December 20, 2006.

wrecking some of the improvements that had been made to date and setting back the recovery efforts. The unfortunate turn of events just stoked his enthusiasm for his new job and the commitment it represented.

Among his most important responsibilities was to coordinate a housing recovery plan: helping residents and would-be residents navigate the financial impediments to rebuilding their homes, building new ones or finding desirable, affordable rental housing. In an isolated parish where property traditionally was passed down from one generation to the next and hardly ever was put up for sale, the limitations of available residential property coupled with the extreme costs involved in building post-Rita/Ike homes to meet stringent new government codes threatened to scare young people away to less vulnerable locales, in Lake Charles or beyond. Ryan was quick to grasp the potential long-term impact of that situation if left unchecked, so programs to develop starter homes with mortgage assistance, to provide financial incentives for elevating homes to new flood-prevention standards, and to create workforce rental housing became some of his top priorities.

"It's not pipe dreams anymore," he said. "Now it's things that are starting to become reality, and so people my age, who haven't had the opportunity to come back, will now have the opportunity to come back. That's the whole key. We need to have people of child-bearing age to come back. We have to have the next generation to step in and take over.

"I think the demeanor of the people is starting to change a little bit. It's going from one of fear and uncertainty to a little bit of hope. That's huge. It's that little bit of hope that keeps me coming every morning. I never have a bad day here."

No one was happier to see Ryan back home in Cameron Parish than his great-grandfather. Macilda had died the previous July 4, suffering a heart attack just days before they were to make the move back to the old home place on Severin Road, named for her father. Soon afterward, Charlie moved into the new double-wide mobile home by himself and went on living his life. At ninety-six, he still rode horseback, still worked the black Angus cattle on his property, still went hunting, still liked his toddy—usually Seagram's, some-times Chivas Regal.

Ryan was already close to his Paw-Paw Charlie. Now the bond between them grew even stronger. Ryan reveled in the opportuni-ties to spend quality time with him, whether taking him to his doc-tor's appointments, driving him around Grand Chenier and Creole to visit friends and relatives, or just sitting at the new kitchen table, sharing a drink and talking. He continued to learn from Charlie's example.

It was about that time that Charlie started talking about passing on. Losing Macilda after seventy-three years of marriage, he had soldiered on, but a year later he was getting more and more pensive about his life, and death. He told Ryan one day, "You know, my boy, if heaven's any better than what I've lived through on this earth, I'll have to see it to believe it."

Ryan sensed that his great-grandfather was content. And several times after his return to Cameron Parish, over glasses of Seagram's and water, Ryan tried to make his Paw-Paw Charlie understand that he not only respected the old man for the life he had led, he envied him, too. Charlie was proud of the young man for going off to col-lege, getting his two diplomas and coming back to work for the par-ish government. He'd tell Ryan, "You're just too smart for me."

Invariably, Ryan would try to convince him that he would gladly trade his book-learning for Charlie's life experience. That's why Ryan always listened so intently whenever the old folks had something to say.

"They suffered and they dealt with the suffering and they overcame it and they became better people," he said of Charlie and Macilda. "There was no way that you could challenge him that would make him nervous.

"There's nothing that they didn't see that I possibly could. You lose your family members, lose your home, lose your church, lose your wife of seventy-three years, lose your own children, your own grandchildren, what else is there? What else can you take from a man? I told him, driving home from the hospital the day Maw-Maw passed away, 'You have to be strong for us, at least for a little while.' And he made a year and a half without her."

Charlie died on March 7, 2009, two months past his ninety-seventh birthday. At his funeral at St. Eugene Catholic Church in Grand Chenier, Ryan delivered the eulogy.

He spoke fondly of his great-grandparents' relationship with each other.

"I can never forget Paw-Paw dragging Maw-Maw down the aisle at Mass but never letting go of her hand and Maw-Maw having no choice but to accelerate almost to a jog as not to fall, but all the while marching up the aisle with a smile on her face," he said. "For seventy-three years they were inseparable. His guardian angel was the peacekeeper of our family, the calming voice, and the only one who had the courage to say, 'Charlie that's enough!' following one too many glasses of whiskey. I think he knew how better a man she made him and after seventy-three years, he still opened the door for her, helped her in and out of the car, would wash the dishes, folded clothes, and when she had a bypass at ninety he never left her side."

And he spoke of how rewarding it was for him, and his friends, to spend quality time with his great-grandfather over the years.

"I truly hope that he realized that it was never the glasses of whiskey that kept bringing people back to visit, but rather the words spoken over those glasses," Ryan said. "My friends were enamored with Paw-Paw, and rightfully so. Whether I needed relationship advice, gardening advice, or a drink, they knew where I would go. It's that relationship with him that I will cherish."

Four months after his Paw-Paw Charlie's funeral, Ryan got married.

The previous summer, just weeks after getting back home and settling into his new job, Ryan had proposed to Megan. They had been together almost three years by that point and knew they were meant to be together, but Megan hadn't anticipated spending their life together in Cameron Parish. She made the leap of faith, though, moved down there, took a teaching job and began working on her teacher certification at McNeese State University in Lake Charles.

"In so many ways, she reminds me of my Maw-Maw 'Cilda," Ryan said. "She really does. She always has a smile on her face, and Maw-Maw was always smiling."

The couple settled in Grand Lake. Ryan joined the Lions Club and the Knights of Columbus. Megan immersed herself in school duties, community projects, and church activities. The product of a rowdy, happy, Italian-American family and a fan of rap music and much of modern-day, mainstream American entertainment and culture, Megan would marvel at times about the way she ended up in the farthest reaches of Louisiana's Cajun coast, happily married to an old young man whose idea of a good time was to take off his shoes, sit on the porch in his rocking chair, listen to some Cajun music—the classic, 1950s-era stuff by Iry Lejeune and Nathan Abshire—have a glass of lemonade, and watch the sun go down. Despite her initial trepidation, she embraced the Cameron Parish way of life, where there's nothing to do, but they're never bored.

"My friends in Baton Rouge said, 'Ryan, what are you thinking, going back? You were set here,'" he said. "Even some of my family members down here said, 'Why in the world would you do that?' You know, there comes a point in time where all the money in the world can't buy you self-gratification."

It's a place that's worth moving back to, he would tell them.

"Most of the people that grew up here were exiled from Nova Scotia in *le grand derangement*. We came to a place where there were Attakapas Indians and that was it. We settled the area. My family has been here since the early 1800s, and they're buried here. That included Charlie and Macilda."

In time, Ryan left the parish government for similar work—hurricane recovery and community planning—with an Abbeville-based consulting firm. He and Megan started a family. And, as if

After Hurricane Rita, Charlie and Macilda Theriot donated this statue of the Blessed Virgin Mary to St. Eugene Catholic Church to replace one donated by her parents, Severin and Lidian Miller, that was lost in Rita's floodwaters.

Ryan Bourriaque at the graves of his great-grandparents Charlie and Macilda Theriot, St. Eugene cemetery, Grand Chenier.

there was ever any doubt, they built a home on the family property, on Severin Road in Grand Chenier.

"I'm the sixth generation on this ridge. That means a lot," Ryan said. "I appreciate that the historical sense is still present here. I fully intend for my children and grandchildren to know a little bit of the Cameron that I knew. It won't be the same, it never will be the same as it was before Rita, but it can't hurt us to try. If we don't try, then it'll never happen."

ACKNOWLEDGMENTS

As a career newspaperman, I long ago grew accustomed to the instant gratification of writing or editing a story one day and seeing it in print the next day. I had never written a book before taking on this project, so the notion of spending years to produce a single published work required some adjustment on my part. While unfamiliar, the experience has been deeply rewarding, and I am indebted to many kind and gracious individuals who helped to make this book possible.

Ever since my editor at *The Times-Picayune* assigned me to write a series of articles taking stock of Louisiana's Cajun culture at the start of the new century, three heroes of modern-day Cajun Louisiana have been my guardian angels. Over the past twelve years they have patiently entertained my many questions, engaged me in provocative and enlightening discussions, and provided valuable support for my assorted journalistic forays into Acadiana, including this one. To Dr. Carl Brasseaux, eminent chronicler of Acadian and Cajun history; Warren Perrin, proud Cajun cheerleader and righter of historical wrongs; and Amanda LaFleur, Cajun French instructor *extraordinaire* at Louisiana State University and my lifelong friend, *merci beaucoup*.

My work was aided by a Louisiana Publishing Initiatives grant from the Louisiana Endowment for the Humanities. Thanks to Dr. Michael Sartisky, LEH director; John Kemp, the *very* helpful administrator of the LEH grant program; Dr. Ray Brassieur, assistant professor of anthropology at the University of Louisiana at Lafayette, who provided a detailed, thoughtful critique of my manuscript as the grant's academic expert; and Susan Larson, reviewer of books and champion of authors, who induced me to investigate the grant

189

program in hopes that I would write a book—any book.

At *The Times-Picayune*, editor Jim Amoss, managing editor Peter Kovacs, and suburban editor Kim Chatelain have long supported my efforts to tell stories such as this one. Photo editor Doug Parker was most helpful in arranging the use of newspaper photographs in the book. The sustained encouragement of my newspaper colleagues, particularly Robert Rhoden, Karen Baker, Ellis Lucia, Scott Lemonier, and David Grunfeld, was more important to me than they realized; I am grateful for their friendship.

Over the course of my research, I was overwhelmed by the hospitality of storm victims willing to welcome a stranger into their homes, offices, or boats and share their stories, usually over a couple cups of coffee. Several good friends pointed me toward many of my interview subjects, including Charles and Lanette Hebert, Dr. Bradley Leger, Ray and Brenda Trahan, Bill Gonsoulin, Maureen Theriot Bergeron, and Natial d'Augereau. From West Virginia, Amanda Hanson provided me with an extensive account of her experiences in Cameron Parish as an AmeriCorps volunteer. For assistance in locating and obtaining photographs, thanks to Vermilion Parish agricultural agent Andrew Granger, Diocese of Lake Charles communications director Morris LeBleu, Dr. Janet Grigsby of Union College, Keith Magill of the *Houma Courier*, and Terry Maillet-Jones of the *Beaumont Enterprise*.

Esteemed geographer/historian Richard Campanella of Tulane University offered sage advice about approaching the University of Louisiana at Lafayette Press with my manuscript. The press was my first choice largely because of the excellent job it did with his *Geographies of New Orleans*, one of my favorite books. I couldn't be more appreciative for the two maps he was kind enough to create to illustrate my narrative.

The good folks at the University of Louisiana at Lafayette Press took great care in shepherding my work to publication. Assistant director James D. Wilson Jr., production assistant Jessica Hornbuckle, and marketing director Melissa Teutsch are great at what they do, and their patience with someone who had no frame of reference for anything having to do with the publishing world was a welcome relief to this rookie author.

Lastly, and mostly, I thank my wife Robyn. Throughout this years-long odyssey, she has been a vital sounding board, an enthu-

siastic travel partner, a skilled proofreader, and a loving and sup-
portive companion. To whatever extent this work is successful, she
should share in the credit.

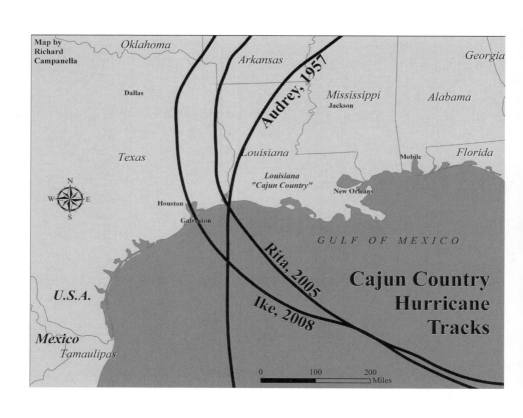

TIMELINE

1605
Acadie is established by the French at modern-day Nova Scotia, Canada.

1636
Pioneer families begin arriving from France to populate the colony and begin to establish the Acadian culture in the New World.

1682
Explorer Rene-Robert Cavalier, Sieur de La Salle, claims Louisiana for France. In the same year, he reports a village of the Houma Indians near what is now West Feliciana Parish.

1718
New Orleans is founded.

1720s-1740s
European settlers begin immigrating to South Louisiana.

1755
As war with France looms, British authorities—who had ruled Nova Scotia since 1713—begin to expel French-speaking, Catholic Acadians.

1762
The Treaty of Fontainebleau transfers control of Louisiana from France to Spain. Spanish newcomers begin to join other settlers in South Louisiana.

1764-1785
Acadian exiles arrive in South Louisiana.

1800
France regains control of Louisiana from Spain via the Treaty of San Ildefonso.

1803
The United States acquires Louisiana from France through the Louisiana Purchase.

1812
Louisiana enters the Union as the eighteenth state.

1822
Terrebonne Parish is created.

1836
Vermilionville, later known as Lafayette, is incorporated.

1844
Vermilion Parish is created.

1848
Houma, named for the local Indian tribe, is incorporated.

1861
Charleston, later known as Lake Charles, is incorporated.

1870
Cameron Parish is created.

1901
Louisiana's oil industry dawns with the first successful oil well near Jennings.

1916
State law requires Louisiana residents to send their children to school.

1918
The Catholic Diocese of Lafayette is established. The southwestern Louisiana territory had been part of the Archdiocese of New Orleans.

1921
A new state constitution bans the use of any language other than English in schools, leading to decades of punishment and ostracism for students from French-speaking families.

1928
Huey Long elected governor. A comprehensive road-building program launched by his administration increases the amount of paved roads in the state from three hundred to five thousand in four years, including many of the first paved roads through Acadiana.

1937
Cameron Parish Courthouse is constructed.

1947
The first offshore oil drilling platforms beyond the sight of land begin operating in the Gulf of Mexico off the Louisiana coast.

1957
Hurricane Audrey strikes Cameron Parish, killing between four hundred and six hundred residents. The bunker-like courthouse was one of the few buildings in the parish to survive the storm.

1964
As Hurricane Hilda makes landfall, a water tower crashes into Erath Town Hall, killing eight men in the civil defense office.

1980
The Diocese of Lake Charles is created.

1990
Census cites 668,271 Americans identified as Acadian or Cajun; most live in Louisiana.

2003
Great Britain's Queen Elizabeth II apologizes for the expulsion of the Acadians from Nova Scotia two and a half centuries earlier.

2005
August 29 – Hurricane Katrina strikes greater New Orleans and the Mississippi Gulf Coast. The Cajundome arena in Lafayette be-

comes a major reception site for Katrina refugees.

September 18 – Tropical Storm Rita becomes the seventeenth named storm of the Atlantic hurricane season.

September 24 – After sweeping past the Louisiana coast and swamping communities throughout lower Terrebonne, St. Mary, Iberia, Vermilion, and Cameron parishes, Hurricane Rita makes landfall as a Category 3 hurricane near the Louisiana/Texas border.

2007
Rev. Glen John Provost of Lafayette is named the third bishop of the Diocese of Lake Charles.

2008
September 1 – Hurricane Gustav comes ashore in Terrebonne Parish. On the same day, Tropical Storm Ike forms far out in the Atlantic Ocean.

September 3 – Ike strengthens to hurricane status.

September 9 – Ike crosses Cuba and enters the Gulf of Mexico, following a west-northwesterly track similar to Rita's.

September 12 – Ike sweeps past Louisiana, causing the worst flooding in recent memory in much of Terrebonne Parish and re-flooding many other Louisiana communities still recovering from Hurricane Rita three years earlier. The eye of the storm makes landfall at Galveston, Texas.

PHOTO CREDITS

Brett Duke, *The Times-Picayune*: 32, 35, 41, 2*

Chris Granger, *The Times-Picayune*: front cover (bottom), 18, 28, 78, 88 (bottom), 105, 181 (top), 1* (bottom), 7*, 8*

David Grunfeld, *The Times-Picayune*: 3, 68, 82, 88 (top), 100, 1* (top), 3* (top), 5*, 6* (top)

Diocese of Lake Charles: 129, 131, 133, 139

Ellis Lucia, *The Times-Picayune*: front cover (top), 26, 51, 3* (bottom), 4*, 6* (bottom)

Helen Sagrera: 151 (top)

Houma Courier: 159 (top), 163 (bottom)

Lauren Guidi: 71

LSU AgCenter, Abbeville: 51, 57

Mark Hancock, *Beaumont Enterprise*: 121, 183

Matt Stamey, *Houma Courier*: 168 (bottom)

Mel Landry: 73

National Oceanographic and Atmospheric Administration: 10

Ron Thibodeaux: 21, 23, 38, 43, 49, 62, 65, 95, 111, 115, 137, 151 (bottom), 159 (bottom), 163 (top), 167 (top), 181 (bottom), 187

*Page found in the photo gallery.

INDEX

RON THIBODEAUX

A writer and editor for more than 30 years at the *New Orleans Times-Picayune*, Ron Thibodeaux has been described as "a noted chronicler of Cajun country."

In travels throughout his native South Louisiana, he has written about Cajun culture in all its aspects, from killer hurricanes to men's supper clubs, from dancehall fiddlers to alligator wranglers, from the uncertain future of Louisiana's native French language to the challenges of keeping the Cajun experience authentic while marketing it to tourists. He also has ventured to the Acadian homeland of Nova Scotia to explore the modern-day similarities and distinctions between Louisiana's gregarious Cajun people and their more reserved Acadian cousins north of the border.

Ron is a graduate of Louisiana State University. He and his wife Robyn live in Covington, Louisiana; they have three children and seven grandchildren.